Climate Responsibility in Business Models

Management, Digital and Technological Perspectives

Adam Jabłoński

CRC Press
Taylor & Francis Group
Boca Raton London New York

CRC Press is an imprint of the
Taylor & Francis Group, an **informa** business
A SCIENCE PUBLISHERS BOOK

First edition published 2025
by CRC Press
2385 NW Executive Center Drive, Suite 320, Boca Raton FL 33431

and by CRC Press
4 Park Square, Milton Park, Abingdon, Oxon, OX14 4RN

CRC Press is an imprint of Taylor & Francis Group, LLC

Library of Congress Cataloging-in-Publication Data (applied for)

ISBN: 978-1-032-74062-1 (hbk)
ISBN: 978-1-032-75533-5 (pbk)
ISBN: 978-1-003-47440-1 (ebk)

DOI: 10.1201/9781003474401

Typeset in Times New Roman
by Prime Publishing Services

I dedicate this book to Professor Henryk Arctowski and Professor Antoni Bolesław Dobrowolski, Polish polar explorers and members of the famous Belgica scientific and research expedition, considered the world's first climate scientists

Preface

The appropriate understanding and shaping of the principles of climate responsibility in the business models of various types of organizations, based on scientific cognition, is a huge challenge of modern science. In order for science to be a weapon of truth, it must be strongly focused on practical application. Then, the decision-making mechanisms implemented by different decision-makers will result in the proper results in the economy, business and society. Such an approach should be used in solving climate problems, which are currently one of the core strategic factors creating crisis phenomena at various levels of human existence along with the dynamic digital, electricity and climate transformation of the surrounding reality. The aim of the monograph is to present the author's original multidimensional approach to understanding, applying and creating the principles of climate responsibility in the context of the development of digital technologies applicable to the business models of companies. The idea and concept of the monograph was inspired by the fact that the author is a member of a research team participating in a scientific project financed as part of the foreign contest entitled European Climate, Infrastructure and Environment Executive Agency (CINEA) CINEA.C – Green research and innovation C.1 – Horizon Europe Climate, GRANT AGREEMENT, Project 101056933 – Climateurope2. This project refers to the fact that the reality of climate change is felt strongly in unprecedented extreme climatic conditions, among many other impacts that have significant consequences for socio-economic systems nowadays. The scientific problem posed in this way creates a wide range of possibilities for a full change in the logic of climate management from the perspective of a deliberate and strongly justified reversal of the mechanisms of adverse and/or neutral impact on the climate and its changes through the dynamic and creative development of climate-changing business models supported by widely dedicated strategies. The concept of positive orientation in management applies here (Luthans et al. 2001, pp. 3–20; Luthans 2002a, pp. 57–72; Luthans 2002b, pp. 695–706; Luthans and Church 2002, pp. 57–75; Luthans and Avolio 2003, pp. 241–261).

The book provides expertise in strategic climate management in relation to the modern principles of digital and climate transformation examined through the concept of business models. This monograph aims to broadly illustrate the complexity of the issue presented through climate responsibility shaped at the level of business, the economy and society. The components of the author's original concept of climate responsibility in business models in management terms and from digital and technological perspectives have been presented graphically in Figure 1.

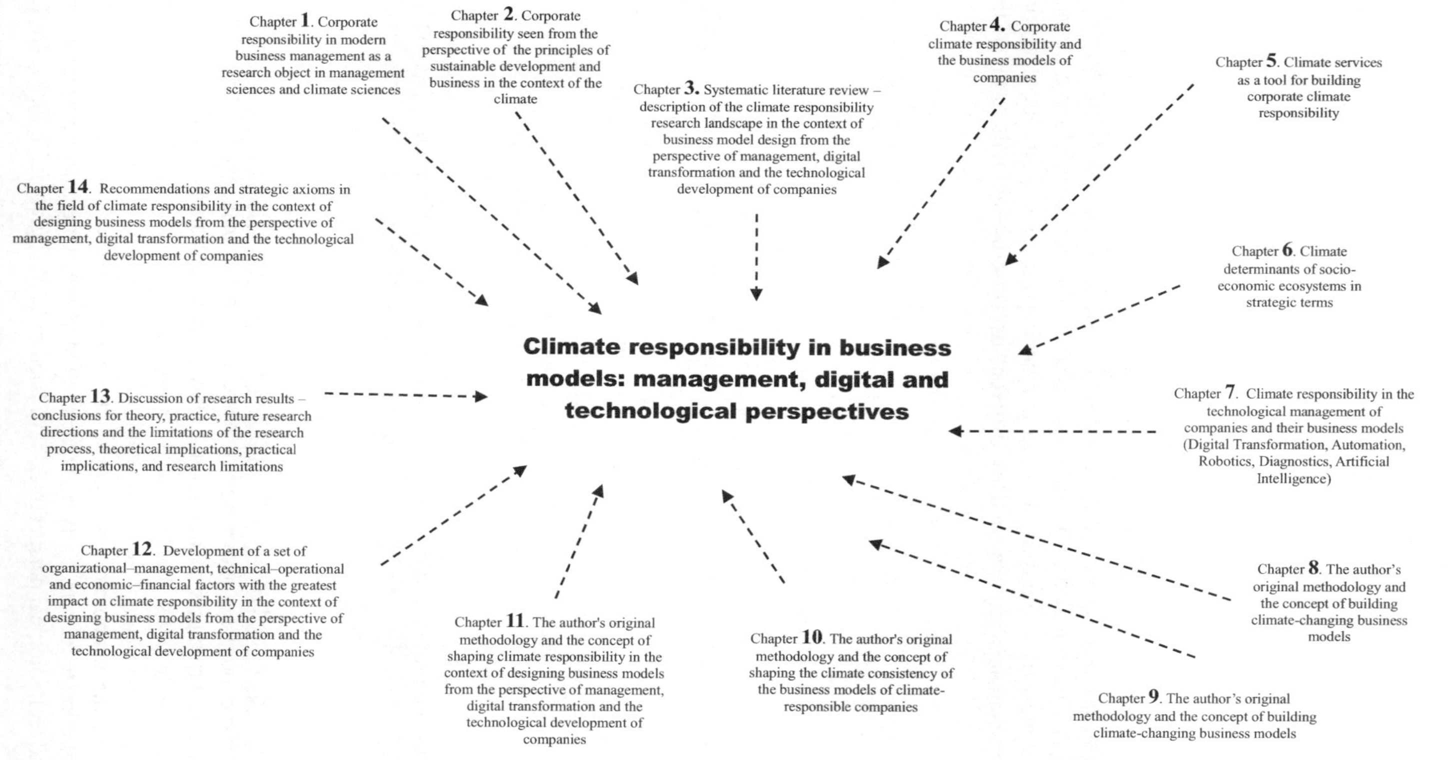

Figure 1. Illustration of the structure of the monograph: Climate responsibility in business models: management, digital and technological perspectives. Source: Own study.

As part of the presented model, a closed cycle of implementing the principles of climate responsibility in business models focused on management, digital and technological perspectives was identified.

This proposed model is described in detail in this scientific monograph. The aim of the monograph is to indicate the theoretical and practical determinants of building and creating the principles of climate responsibility embedded in business models by developing management mechanisms open to digital and technological perspectives. The book is preceded by this preface.

The first chapter broadly describes the principles of corporate responsibility in modern business management as an object of research in management sciences and climate sciences. The second chapter refers to the principles of corporate responsibility seen from the perspective of the principles of sustainable development and business in the context of the climate. The aspects of the systematic literature review concerning the description of the climate responsibility research landscape in the context of business model design from the perspective of management, digital transformation and the technological development of companies are described in chapter three. In turn, the fourth chapter discusses the assumptions and principles of corporate climate responsibility against the background of the concept of company business models. Chapter five deals with climate services as a tool for building corporate climate responsibility. Chapter six presents the climate determinants of socio-economic ecosystems in terms of strategy. Chapter seven presents the principles of climate responsibility in the technological management of companies and their business models (Digital Transformation, Automation, Robotics, Diagnostics, Artificial Intelligence). Four subsequent chapters, i.e., chapters eight, nine, ten and eleven, present the author's original methodologies and concepts for developing the principles of corporate climate responsibility. Chapter eight describes the author's original methodology and the concept of building climate-changing business models. Chapter nine presents the author's original methodology and the concept of shaping the climate maturity of the business models of climate-responsible companies. In turn, chapter ten describes the methodology and the concept of shaping climate consistency in the business models of climate-responsible companies, and chapter eleven describes the methodology and the concept of shaping climate responsibility in the context of designing business models from the perspective of management, digital transformation and the technological development of companies. Chapters twelve, thirteen and fourteen are recommendation chapters on shaping corporate climate responsibility in relation to business models. Chapter twelve concerns the development of a set of organizational–management, technical–operational and economic–financial factors with the greatest impact on climate responsibility in the context of designing business models from the perspective of management, digital transformation and the technological development of companies. Chapter thirteen presents the discussion of research results–conclusions for theory, practice, future research directions and the limitations of the research process, theoretical implications, practical implications, potentially interesting directions for future research, and research limitations. Chapter fourteen refers to recommendations and strategic axioms in the field of climate responsibility in the context of designing business models from the perspective of management, digital transformation and the technological development of companies. The monograph ends with the conclusion and a list of tables and figures.

The main objective of the scientific discussion

The main objective of the scientific discussion in this monograph is to present the author's original theoretical and practical solutions for shaping the principles of climate responsibility from the perspective of the business models of companies embedded in the logic of digital and climate transformation at the level of business, the economy and society.

Setting a scientific problem and identifying a cognitive gap

As part of the scientific goal, it is crucial to determine the scientific problem and identify the cognitive gap. This cognitive gap is an identifiable set of factors shaping the principles of climate responsibility in business models through effective and efficient management mechanisms open to and focused on digital and technological perspectives.

Such a defined cognitive scope should lead to the emergence of a new concept of the interpretative development of climate responsibility in business models through management as well as digital and technological perspectives.

This concept should fit into the current conditions of the global approach to the development of climate responsibility in business models. Theoretical and practical factors determining the optimal shaping of the principles of climate responsibility for effective and efficient management supported by the digital and climate transition should emerge when set against the background of this concept.

The author of the monograph understands climate responsibility in business models as an interdisciplinary set of approaches to the development of sensitivity and climate awareness among managers and entrepreneurs, who are convinced of the opportunities for development and growth of companies that they manage resulting from opportunities embedded in the potential of climate threats. Such a reversed approach and the adopted polarization of the perception of climate change by managers and entrepreneurs is a solution resulting from creative strategic thinking based on scientific cognition supported by business experience.

This approach to research is part of a broad approach to shaping the concept of climate responsibility in business models open to digital and climate transformation against the background of sciences such as economics and finance, climate sciences, and management and quality sciences. The presentation of the issue in this way requires the definition of new assumptions of the theory of climate responsibility in business models.

Background to the scientific discussion

As part of the adopted theoretical assumptions related to the solution of a defined scientific problem pertaining to climate responsibility in business models, an analysis was carried out to embed this problem in the theory of management sciences. In order to correctly diagnose the phenomenon occurring in science, it is important to refer to scientific discussions determining the principles of the adopted, appropriate interpretation. For this purpose, however, it is necessary to combine the determinants of understanding climate responsibility with the concept of business models in the strategic management of companies against the background of shaping the mechanisms of digital and climate transformation.

The adopted research methodology and the scope of research and scientific discussion

As part of the research method for building the principles of climate responsibility in business models, a sequence of activities, which is presented in the Table 1, has been adopted.

Table 1. Sequence of activities within the research method of climate responsibility in business models.

No.	Sequence of activities within the research method of mechanisms of climate responsibility in business models
1.	Collecting relevant literature on the subject of climate responsibility in business models.
2.	Identifying the components of the climate responsibility model in business models.
3.	Defining the scope of influence of various factors on the development of climate responsibility in business models.
4.	Choosing solutions to explore climate responsibility in business models.
5.	Defining the features characterizing the specificity of creating climate responsibility in business models.
6.	Analyzing the contextual conditions of the market in relation to the development of the principles of climate responsibility in business models and their practical implementation.
7.	Defining good practices in shaping climate responsibility in business models.
8.	Developing analyses, the resulting synthesis, conclusions and recommendations for building climate responsibility in business models.
9.	Determining the theoretical determinants of creating climate responsibility in business models.
10.	Indicating new directions of development for climate responsibility in business models in theoretical and practical terms.

Source: Own study.

The diagram of the process of the research procedure in empirical sciences and the process of diagnostic procedure in psychology developed by Spendel (2005, p. 33) interpreted for the purpose of conducting advanced research in the area of climate responsibility in business models is presented in the Table 2.

The crucial conditions necessary to maintain the correct implementation of the research strategy have been maintained. This enabled the creation of both a research framework and the conducting of extensive theoretical and practical research. The study is a cognitive–evaluating study based on multidimensional scientific inference. The book covers both the assumptions of creating a new theory of climate responsibility in business models and elements covering practical solutions.

Table 2. Diagram of the process of the research procedure in empirical sciences and the process of diagnostic procedure in psychology developed by Spendel, Z. (2005) interpreted for the purpose of conducting advanced research in the area of climate responsibility in business models

Scientific research – classical	Scientific study on climate responsibility in business models		Diagnostic procedure – classical	Diagnostic procedure – climate responsibility in business models
Recognizing and verbalizing the research problem concerning the relationship between selected phenomena, formulating a preliminary hypothesis and (possibly) showing the circumstances of the problem	The recognition and verbalization of the research problem results from the current needs of the manager and he/she is the generator of the research problem. The research problem is a reflection of a real managerial problem in a business run by a manager. He/she can independently conduct the research process with regard to the research problem posed and/or use the help of a professional researcher. The circumstances of addressing a given research problem are the specific, actual and current context of the organization in a given socio-economic ecosystem in which the company exists.	**Conceptualization**	Preliminary identification and assessment of the problem and justification for addressing it – usually it means finding some irregularities	A problem is recognized and verbalized as part of a diagnostic procedure only by a manager who needs to solve a given diagnostic problem. The manager, noticing some problems that occur in the organization, looks for effective and efficient ways to solve them.
Positioning the problem and hypothesis in a theoretical context, defining the assumptions and formulating research subhypotheses in terms of relationships between the separated variables	The manager and/or the researcher who supports him/her quantifies and specifies his/her problem. They are an input to the specification of the adopted research problem, which is in fact a business problem because it directly results from the business needs of a specific company.	**Specification**	Formulation of detailed questions and diagnostic hypotheses on the genesis and pathomechanism of the identified abnormality	The manager develops a problem that bothers him/her in detail, which he/she must solve. For this purpose, it generates research questions or hypotheses. .
Determination of the criteria of empirical applicability and quantitative recognition for observational terms occurring in subhypotheses	The manager and/or the researcher assisting him/her develops an execution framework in terms of the algorithm of solving a given problem. He/she can use artificial intelligence solutions here.	**Operationalization**	Planning (or selecting an appropriate model) of the diagnostic procedure and the selection of suitable (i.e., relevant to the hypotheses) diagnostic methods	The manager prepares a list of specific steps along with the proposed diagnostic methods.

Selection of models of collecting and organizing experience compared with the content of empirical implications of hypotheses and models of data analysis. Application of selected research and statistical models	The manager and/or the researcher assisting him/her makes a constructive comparison of the principles of experience management in the organization and the results of evidence research, comparing the results of resource management in a given organization with the research problem to be solved.	**Comparison**	Collection of data by means of selected methods and analysis of results in terms of their compliance with diagnostic hypotheses	The manager aggregates the obtained results and compares them with a problem in the area of diagnostic procedures, verifying them through previous decision-making processes in the same or a similar area of business.
Formulation of conclusions on the veracity of the tested hypotheses, as well as the indication of the acceptable reference range of the obtained research results to: 1) assumptions and research hypotheses, 2) the possibility of formulating further problems, possibly proposing theoretical and methodological postulates, revising the current paradigm, and 3) taking practical actions on that basis	The manager and/or researcher assisting him/her will develop a set of goals to achieve research results. It takes into account, in the form of possible micro-corrections, information, data and knowledge from other sources supporting the decision-making processes.	**Application**	Formulation of the diagnosis (in the form of conclusions, opinions, rulings, etc.) and postulates regarding corrective, therapeutic conduct and cooperation with other specialists and institutions	The manager interprets, diagnoses and synthesizes the results obtained as a result of the procedure, correcting them, if necessary, for other resulting sources that may correct potential errors in the decision-making diagnostic process.

Source: Own study based on Spendel (2005, p. 33).

The subject of the scientific discussion

The subject of the scientific discussion includes scientific and practical digressions on components forming a coherent system of mechanisms of climate responsibility in business models.

Cognitive objectives

Cognitive objectives include:

1. Identifying the factors responsible for shaping climate responsibility in business models.
2. Determining the relationships between individual factors and attributes in shaping the principles of climate responsibility in business models.
3. Assessing the impact of climate responsibility in business models on the effectiveness and efficiency of organizational management.
4. Assessing the relationship between the principles of climate responsibility and the construction of business models that positively influence climate change from the perspective of dynamic processes of digital and climate transformation.
5. Assessing mechanisms for shaping the maturity and climate cohesion of business models of companies as a key to the development of the principles of corporate climate responsibility.

Methodological objectives

Methodological objectives include:

1. Indicating the configuration of factors that shape the principles of climate responsibility in business models.
2. Developing strategic recommendations for the development of the principles of climate responsibility in business models.

Utilitarian objectives

Utilitarian objectives include:

1. Developing the author's original methodology for shaping climate responsibility in business models.
2. Developing a set of management, technical and operational factors with the greatest impact on creating the principles of climate responsibility in business models.

Key research questions within the framework of qualitative research

The following research questions were posed within the framework of the defined research objectives:

1. What is the place and role of climate responsibility in building business models that positively change the climate?
2. How do digital transformation and climate transformation affect the design and implementation of business models that meet the criteria of climate responsibility?

3. What should the interpretation and configuration of business models that positively change the climate be?
4. How should we understand the principles of climate responsibility and what attributes should be used to describe this in the space of digital and climate change?
5. What climate management mechanisms in business models support the digital transformation of climate-responsible companies?
6. How can mechanisms for building the climate maturity of the business models of climate-responsible companies be shaped?
7. How can the mechanisms of climate cohesion of the business models of climate-responsible companies be shaped?

Embedding the topic in scientific theories:

Due to the multidimensionality and holistic nature of the functioning of companies, the topic of the scientific discussion requires one to address issues related to the theory of management and quality science, economics and finance, climate science and the theory of technical sciences.

This book is based on the following assumptions:

Macroeconomic assumptions

1. As far as the principles of digital transformation and climate transformation are concerned, they are definitely global in nature, of course, with varying degrees of intensity conditioned by social, technological and organizational maturity.
2. The assumptions and principles of climate responsibility serve to stimulate climate awareness and sensitivity at the level of business solutions, often of a universal nature, and thus are generally available.
3. Global climate, social, health, ecological, political, military, economic and financial crises adversely affect the development and formation of climate responsibility and should be strongly taken into account in decision-making in various fields of economic and social decisions as well as in the implementation of investment projects of various dimensions and significance.

Sectoral assumptions

1. Sectors of the economy, including organizations of various types, based and functioning on the joint implementation of technical, operational, organizational, management and economic–financial aspects undergo dynamic transformation processes to meet the challenges of the digital and climate economy, which generates the creation of new business models that break the current rules in the sector in an innovative way and meet the principles and criteria for the concept of a "game changer".
2. The professionalization and specialization of sectors and industries are crucial, as are the expectations and requirements of societies, including local ones, which forces the creation of social and climate enterprises where, in addition to economic profit, it is important to generate social and climate profit resulting from the development of intelligent climate values.

3. The standardization of the design and implementation of business model development processes requires consideration of the formal and legal conditions relevant to companies embedded in specific sectors and industries, including legal and economic issues as regulatory components at the level of business, the economy and society.
4. Businesses and various types of organizations operating in sectors and industries exist by complying with the relevant laws that underlie the applicable regulations for their activities in numerous areas and dimensions, especially in the area of ecological and climate criteria in order to reduce their adverse impact on the climate and the environment. This has a significant impact on the determinants of designing effective and efficient business models dedicated to them within the sectoral conditions in which a given company operates and the implemented principles of digital transformation and climate transformation at the level of given sectors and industries of the economy.

Microeconomic assumptions

1. A large number of companies which have a notably adverse impact on the climate and the environment are not fully aware of how widely they affect and generate far-reaching and often long-term climate and ecological losses in the natural ecosystem. This lack of thorough knowledge generates a number of problems at the level of their functioning in both the short and long term.
2. The business, economic and social environment forces dynamic changes in business models and their configuration at the company level, especially in terms of their adverse impact on the climate and the environment.
3. Modern business models and digital and climate transformation solutions are transforming business and the economy into a set of digitally and technologically supported solutions to maintain business continuity and the ability to survive even in the most difficult conditions, along with a dynamic revolutionary and evolutionary process adapting to the new determinants of business and economic development and the behavior of societies, including local societies.

This monograph is based on the following limitations

1. So far, there has been no such extensive research in the field of a multidimensional approach to understanding, applying and creating the principles of climate responsibility in the context of the development of digital technologies applicable to the business models of companies. There is not a great deal of experience in this field in the wide stream of scientific publications.
2. The multitude of approaches and interpretations to naming and codifying different types of business models makes it difficult to establish a single binding hierarchy (taxonomy) of business models in this area, especially taking into account the principles of climate responsibility.
3. The presented examples of interpreting and naming climate responsibility in the context of the criterion of digitalization and climatology principles characterize

different sectors of the economy, which on the one hand is a limitation, but the unification and broad interpretation of such a theoretical and practical description of climate responsibility can be extremely valuable for various criteria and attributes and characteristics specified in various sectors of the economy.

Time relevance, the importance of issues and the contribution to science of issues related to socio-technical systems in the formation and development of digital business models

By writing this monograph, the author wished to bring added value to the identification, conceptualization and operationalization of the implementation of mechanisms for understanding, applying and creating the principles of climate responsibility in the context of the development of digital technologies applicable to the business models of companies.

The proposed research issue is generally not presented in such a broad and multidimensional approach and interpretation. The value of the author's original, creative contribution is related to the presentation of a comprehensive approach to understanding, applying and creating the principles of climate responsibility in the context of the development of digital technologies applicable to the business models of companies and also in the creation of innovative business solutions in this area. It also takes into account operational aspects within the framework of the implemented process and design excellence of the organization. The author also wished to draw attention to the need to build a new theory of building and implementing the principles of climate responsibility, climate maturity and climate cohesion in business models focused on the digital and technological development of business, the economy and society, and to indicate utilitarian solutions that provide guidance to managers of various types of companies interested in developing the principles of climate responsibility in business models.

Qualitative research and case studies

The main objective of the scientific research in this monograph is to demonstrate and present the author's original, multidimensional approach to understanding, applying and creating the principles of climate responsibility, climate maturity and climate cohesion in the context of the development of digital technologies applicable to the business models of companies. The book includes extensive literature studies based on a systematic and critical literature review and modelling based on the analysis of two case studies, namely:

1. A case study of the railway sector presenting selected issues and specific solutions related to the construction of hazard registers in rail transport, extended to incorporate elements related to climate and digital threats.
2. A case study of the water and sewage sector presenting selected aspects of the design and implementation of climate-changing business models used in companies in the water and sewage sector. The scope of the chapter includes the identification of selected organizational and management solutions built in business models which change the climate and are applicable in the water and sewage sector.

In addition, the author's original scientific solution was proposed in two theoretical and practical spheres, namely:

1. Presentation of the author's original multidimensional approach to understanding, applying and creating the principles of climate responsibility in the context of the development of digital technologies applicable to the business models of companies.
2. Presentation of the author's original multidimensional approach to understanding, applying and creating the principles of climate maturity in the context of the development of digital technologies applicable to the business models of companies.
3. Presentation of the author's original multidimensional approach to understanding, applying and creating the principles of climate cohesion in the context of the development of digital technologies applicable to the business models of companies.
4. Presentation of the author's multidimensional approach to understanding, applying and creating positively changing business models.

Such approaches can be useful for theorists and practitioners of climate management, and not only in terms of a better understanding of the place and role of climate responsibility in the processes of shaping business models supported by digital transformation and climate transformation solutions.

Finally, the author's original canvas of business models, which positively changes the climate and functioning in the space of revolutionary and evolutionary digital and climate changes, was also developed.

References

Luthans, F. (2002a). Positive organizational behavior: Developing and managing psychological strengths. Academy of Management Executive, 16(1): 57–72.

Luthans, F. (2002b). The need for and meaning of positive organizational behavior. Journal of Organizational Behavior, 23: pp.695-706.

Luthans, F. and Avolio, B.J. (2003). Authentic leadership: A positive developmental approach. pp. 241–261. *In*: Cameron, K.S., Dutton, J.E. and Quinn, R.E. (Eds.). Positive organizational scholarship: Foundations of a new discipline. San Francisco, CA: Barrett-Koehler.

Luthans, F. and Church, A. (2002). Positive organizational behavior: developing and managing psychological strengths. Academy of Management Executive, 16(1): 57–75.

Luthans, F., Luthans, K., Hodgetts, R. and Luthans, B. (2001). Positive approach to leadership (PAL): Implications for today's organizations. Journal of Leadership Studies, 8(2): 3–20.

Spendel, Z. (2005). Metodologia badań psychologicznych jako forma świadomości historycznej [Methodology of psychological research as a form of historical awareness]. Katowice. Wydawnictwo Uniwersytetu Śląskiego, p. 33.

Contents

List of Tables

List of Figures

1

Corporate Responsibility in Modern Business Management as a Research Object in Management Sciences and Climate Sciences

1.1 Introduction

The aim of this chapter is to present crucial issues related to the concept of corporate responsibility in the scientific and practical space from an institutional and individual perspective. The scope of the chapter covers the issue of corporate social responsibility in management processes, as a strategic obligation of each business entity seen from the perspective of stakeholders and business network actors. The subject of the chapter concerns the definition of the general principles of corporate social responsibility in relation to the business, economic and social context.

1.2 Taking responsibility in management processes as a result of the proven competencies of managers creating their business models

The concept of responsibility can be interpreted, applied and explained in different ways. One thing is certain—it is one of the core concepts that determines individual and institutional behavior. Corporate responsibility is one of the crucial attributes and features of decision-making processes at the level of companies and their direct effects. It expresses specific and easy-to-assess actions at the levels of organization and management functions. It determines values and shapes attitudes in institutional and personal terms. It is an expression of concern for the company and its employees. Corporate social responsibility (CSR) expresses a fundamental moral principle about how a business behaves towards society. A company is more than profit. It is the way it works, the foundation that creates morality in the way it approaches not only the customer, but society as a whole. Entrepreneurs have a duty to be aware of the

DOI: 10.1201/9781003474401-1

expectations, goals and values of society (Bowen 1953). Corporate responsibility has numerous dimensions and interpretations. Being responsible is a crucial requirement for managers. A manager has to be responsible for something, for someone, before someone. A manager can also be held accountable. Corporate responsibility at the company level should be considered strategically. It should also be remembered that the responsibility of the organization results from the responsibility of the individuals (people) of whom it is comprised. A.B. Carroll distinguishes four levels of responsibility, namely economic, legal, ethical and philanthropic (Carroll 1991, pp. 39–48). The most commonly used terms and concepts related to corporate responsibility include corporate social responsibility, corporate sustainability management, business responsibility, corporate citizenship, global business citizenship, corporate social performance, corporate social responsiveness, social responsibility, corporate responsibility, community relations, ethical responsibility and ethical management. Companies pursue corporate social responsibility in various and individual ways. There is no uniform recipe for efficiency in the implementation of corporate social responsibility principles at the company level. Therefore, it is extremely important to understand the power of the positive impact of the application of the principles of corporate social responsibility in the logic of practicing management methods and concepts. It should also be noted that the greed of entrepreneurs nowadays is actually "passé". Greed was often implemented in the principles of the value-based management concept of aggressively striving for permanent maximization of the company's profit at all costs by all possible, available means, not always in accordance with the law. The concept of value-based management in metaphorical terms may constitute a "fire" which should be doused with "water", i.e., the principles of corporate social responsibility. Such an interpretation indicates the complexity of the problem related to corporate social responsibility against the background of this concept itself as well as in connection with other concepts currently applicable in modern business management. It should be noted that this concept is particularly important in view of the permanent impoverishment of society, the financial limitations of the so-called middle class and, as a consequence, the constantly decreasing purchasing power of customers that form functioning local communities. The generated symbiosis of relationships between society and business is a substantive and emotional bridge between these groups, and corporate social responsibility is precisely this factor that binds them together. The definition of criteria fulfilling the assumptions of corporate social responsibility that should be followed by the manager of a given organization is currently a moral and ethical priority of intellectually mature organizations. Those criteria are then extrapolated from the strategic level of the company to the tactical and operational levels. Then, all core functions of the organization at these levels are fulfilled and sealed in order to meet and shape the culture of the organization embedded in the principles of corporate social responsibility.

The organizational and management mechanisms defined and adopted for use at the company level have an institutional and individual dimension. Regardless of which of them we choose, the most important role is always played by a person, a manager who often makes managerial decisions guided by the good of the organization based on his/her own, individually accepted and professed attributes of honesty and morality. The assumption of responsibility for the management of the organization must be confirmed by competencies and skills. Competencies are a special and unique combination of education, training, skills, experience and individual psychophysical and psychotechnical characteristics. The individualization of competencies relevant to the assumed and, above all, understood responsibilities and powers is and should be the core attribute of the awareness of professional managers. This results in a specific ability of managers to achieve the assumed results at every level of the organization. These competencies should be developed in a space full of awareness and social sensitivity. It is worth noting that very often managers who assume responsibility for the management of the company do not realize this responsibility and the potential consequences of poor management. This is often due to classic ignorance and excessive personal ambition, unfortunately not supported by the competency required to occupy a high position in business. One should be aware at this point that not having the appropriate competencies and yet taking responsibility for the management of the company results in the poor performance of the organization and thus can already be said to affect all groups of stakeholders centered around an improperly managed company. Thus, responsibility for the business must be dependent on the managerial competencies input and verified throughout the decision-making process as part of the performance of managerial functions and then approved or rejected in the form of an appropriate final vote of approval. This responsibility should be embedded in the business models designed, implemented and operated by managers, both analog and digital ones, where the key to market success is the value proposition offered to customers, relevant to market expectations, along with the wisely created logic of generating income by the company. The value allocation includes the analog and virtual worlds. In this sense, the broadly understood responsibility of managers should be strongly embedded in the business model and strategy of the organization and reflected in the decision-making processes aimed at the implementation of processes and projects related to the allocation of an appropriate value proposition on the market and a responsible and economically and organizationally justified income generation logic.

1.3 Corporate social responsibility in management processes and projects that are at the service of business models and strategies of companies

Business management processes should be based on values expressed by the company in line with the expectations of the surrounding communities. Managers have a range of management concepts, tools, methods and techniques for decision-making and management at the strategic, tactical and operational levels. Their skillful selection determines the market success or failure of the organization. Their application is due to the place and role of the organization in the socio-economic ecosystem. Society currently plays a fundamental role in the space of modern business. Business and society should function in appropriate symbiosis, symmetry and synergy understood in the same way by both interest groups. Society, on the one hand, provides personnel to individual organizations and, on the other hand, is the recipient of the values generated by these organizations. In this case, the business–

economy relationship shapes the mechanisms for defining responsibility in various dimensions, and unfortunately also irresponsibility. Hence, the responsibility of managers is measured by current and future decision-making processes and their effects in the strategic and tactical–operational dimension. These decision-making processes should be carried out in an analysis–synthesis–decision system. In this case, the margin for decision-making error is kept to a minimum. Such operating principles should be embedded in the logic of effective and efficient business management. Corporate social responsibility (CSR) is defined in the European Commission's Green Paper of 18 July 2001 as "the idea that companies voluntarily integrate social and environmental considerations into their commercial activities and in their interactions with their stakeholders" (Green Paper on Promoting a European Framework for Corporate Social Responsibility 2001). The World Business Council on Sustainable Development defines corporate social responsibility (CSR) as "a continuous commitment by business to ethical behavior and to contribute to economic development, while improving the quality of life of the workforce and its families, as well as of the local community and society as a whole" (WBCSD Stakeholder Dialogue on CSR 1998). The implemented mechanisms of corporate social responsibility examined from a strategic perspective are and often become a core competence of the organization. Core competence is defined as a set of diverse skills, skills and complementary resources that form the basis for competing and achieving competitive advantage in a specific industry (Teece and Pisano 1994). Circular processes and unique projects that meet social criteria are at the service of the business model and pro-social strategies. Then, the synergy effect is a platform for strategic change by responsible companies.

1.4 Social responsibility as a strategic obligation of companies and social non-responsibility

Corporate responsibility is embedded at the strategic level of organizations of various types. The strategic pillar shapes the principles of management within the organization and in confrontation with the environment around it. Corporate social responsibility should be the responsibility of every entrepreneur and manager. This awareness should be built into their thought processes strategically, tactically and operationally. This then provides the basis for the company's ability to undertake long-term strategic reflection towards persistence in the market, even in its most difficult conditions. The responsibility of managers directly translates into the responsibility of other members of the organization. This also applies to irresponsibility. These relationships are, for example, as follows in terms of the direct relationship between the employer and the employee:

1. A responsible employer versus a responsible employee.
2. An irresponsible employer versus a responsible employee.
3. A responsible employer versus an irresponsible employee.
4. An irresponsible employer versus an irresponsible employee.

It is socially irresponsible to decide to accept an alternative that the decision-maker considers inferior to another alternative, given the implications for all parties. This usually involves one side profiting at the expense of the whole system (Scott 1977, pp. 185–213). Managers must recognize that investing more time and resources in CSR does not necessarily reduce the likelihood of CSI. Avoiding CSI requires special attention from higher levels and specific resources. The lack of such special attention can have enormous consequences for a significant number of relevant stakeholders (Iborra and Riera 2022, pp. 1–19). The concepts of responsibility and accountability are often confused, although they are two different concepts, and many dictionaries define one in relation to the other. It seems that this was a problem that aroused little academic interest. These concepts have an impact on both project management and general management. The ambiguity of definitions can cause great difficulties for practitioners in allocating "universal" accountability and may cause confusion in the application of management models (McGrath and Whitty 2018, pp. 687–707). The concept of irresponsibility has a negative meaning, but it can be expressed in various interpretations as well as the real effects resulting from it. Often this term is used or heavily abused without formulating the potential of threats resulting from it and extrapolated to frequently occurring broad social and business references. Therefore, responsibility cannot be understood without being aware of the consequences of irresponsibility.

1.5 Stakeholders and/or actors of the business network included in the business model and strategy of the responsible organization

Every company is always composed of people. People from various interest groups are also the recipients of various types of values generated by this company. They are identified as the organization's stakeholders and/or network actors around the organization. As regards stakeholder concepts, their potential to create business value is not often properly understood and valued. The forerunner of stakeholder theory was R.E. Freeman, who defined the concept of stakeholders and indicated the core relationships between the company and the environment. Stakeholders are, in his view, "*any entity (e.g., individual or group) who can influence or is influenced by mission-driven organizations (i.e., set out to accomplish objectives*" (Freeman 1984). The management of the proper structure of stakeholders around the organization is a very important strategic factor in business management. The determinants of achieving positive results in relations with the company's stakeholders are to properly define their contribution to the development and growth of the organization and to dedicate appropriate incentives to them that are relevant to this contribution. Stakeholders are defined as individuals or groups of individuals, organizations, institutions or commercial entities that have a direct or indirect interest in the activities of the company in pursuit of its objectives, which may also affect or be affected by the company (Clarkson 1995, p. 106). Stakeholders interact with the company as well as with each other and can have a real impact on what decisions are made therein (Friedman and Miles 2002). Then, their mutual engagement is crucial. Stakeholder engagement refers to an interaction between a company and its stakeholders that solves knowledge-related problems to improve correspondence in understanding between managers and stakeholders and thus help solve the ethical challenges managers face (Mitchell et al. 2022, pp. 75–94). This specific engagement can also be defined as "a process that creates a dynamic context of interaction, mutual respect, dialogue and change, rather than unilateral stakeholder management" (Manetti and Toccafondi 2012, pp. 363–377). Traditional business theory implies that the orientations of "exchange" and "restrictions" justify the relationship between corporations and stakeholders because, on the one hand, stakeholders exhibit restrictiveness towards corporate entities and, on the other hand, they take value in return (Tokoro 2007, pp. 143–162). Such an exchange of values should be transparent and fair in both the short and long term and contribute to the well-being of all stakeholder groups in the reference period adopted. With regard to the socio-economic ecosystem, we are also dealing with the concept of actor–network theory (ANT), understood as a theoretical and methodological concept in the field of social sciences, derived from constructivist assumptions and based on the achievements of studies in science and technology and the sociology of scientific knowledge (Latour 2005, p. 238). The network effect is created as a result of positive relationships between various types of actors in the business network centered around the organization. From this perspective, the company becomes the so-called node of the network, as a full member in the interactive connections of a specific type of a system, which this network is. In this approach, there is no division in the organization into the so-called internal and external environments; instead, everything is a compact monolith of mutual interaction creating a sum of exchangeable values. This interchangeability is based on mutual responsibility and co-responsibility, and people are the key actors in this network.

1.6 Corporate social performance (CSP) models

The relationship between business and society is very complex and often ambiguous. Whether these relationships are positive, neutral or negative depends on the organization and its orderly process of wise decisions, systemically understood and made, resulting from management processes. The result of this action is an effect aimed at the mutual perception of benefits or losses. The organization's efficiency approach may therefore have a point range closed only in the benefits for the organization's shareholders or a spatial range, where the company is the giver and subject of interchangeability of all values for a wide range of stakeholders and/or actors of the organization's network. A core parameter for measuring the effectiveness of an organization focused on increasing the value of the company to shareholders is the level of results generated by them, measured in financial terms through an increase in revenues, an increase in profits, an increase in fixed and current assets, the cost of capital, tax efficiency and the competitive advantage achieved (Rappaport 1986). However, the performance of companies can be both economic and non-economic. It all depends on the concept of entrepreneurial orientation (EO), which describes the strategic behaviors of companies that materialize in the fact that they are focused on implementing innovations, taking business risks and proactive actions that are more intense and stronger than those of their competitors. It therefore characterizes entrepreneurial behavior at the level of the company as a whole that reflects the decisions made by managers (Wiklund and Shepherd 2003, pp. 71–91). When conducting a complex analysis and resulting synthesis of social responsibility from the perspective of increasing the efficiency of the organization, it is necessary to cite the statements made by Don J. Wood in 1991 (Wood 1991). Corporate social responsibility (CSR) issues have been debated in the USA for decades, but there is no universally accepted answer to the question "Who is a corporation responsible to?" In his opinion, this question has not been answered because this is not the right question. Instead, the question "How can and do corporations contribute to building a 'good society'?" should be asked. This question itself needs to be answered in the context of discussions about corporate social performance. Moreover, to his mind, no productive discussion can take place until our collective thinking about a few key points of language and perception has changed. The main points in his statement are as follows:

1. Neither the old term "business in society" nor the newer term "business and society" properly express contemporary business–society relationships. Corporate social performance is a more comprehensive and descriptive term (Stahel 2006, p. 13).
2. Contrary to many current views, we do not have to choose between the demands of economics and the demands of ethics; ethics is also not something that is attached to economics, as in the phrase "economics and ethics". Economics is ethics, but ethics is more than economics.
3. The cultural relativity of ethics and social control does not absolve us of the responsibility of our own cultures.
4. Responsibility, defined by ethics and enforced by social control, is what makes people and organizations free.
5. Free people (and organizations) choose; they are not forced. If we change our thinking on these points, perhaps we can stop asking so many tangled questions about responsibility and start asking more specific questions about results: Who is harmed and who benefits from corporate actions? Is it fair or just that these harms and benefits should exist? What can be done to reduce harm and increase the benefits of corporate behavior? Then we will come closer to answering the question of corporate social responsibility, and in this process we will also move towards a "good society" and a better world. Everything in this case depends on the environment of a given corporation, which must be properly managed (Miles 1987). There are several models of social enterprises. Yunus was the forerunner of the definition of a social enterprise and its business model. The social business model consists of four components: (1) value proposition (products and/or services addressed to both customers and stakeholders), (2) the balance of social benefits consisting of social and environmental benefits, (3) the constellation of values, including internal and external value chains, and (4) the balance of economic benefits, i.e., sales revenue, cost structure and capital employed (Yunus et al. 2010, pp. 308–325). Social enterprises are therefore hybrid companies that simultaneously pursue a social mission and commercial performance (Wilson and Post 2013, pp. 715–737). This creates unique social

entrepreneurship as "an innovative and socially value-creating activity that can take place within or between non-profit, business or government sectors" (Austin et al. 2006, pp. 1–22).

Such activities contribute to the creation of a special type of organization, i.e., a high-performance organization. At this point, it is worth recalling the definition of a high-performance organization. A high-performance organization is an organization that achieves financial results that are better than those of its peer group over a longer period of time, by being able to adapt well to changes and react to them quickly, by managing for the long term, by setting up an integrated and aligned management structure, by continuously improving its core capabilities and by truly treating the employees as its main asset (de Waal 2007, pp. 179–185). Adding a social component to this definition allows for a strong enrichment of the core attributes of building an organization's competitive advantage and its wider stream of market and social impact.

Models of social enterprises are presented in the Table 3.

Table 3. Models of social enterprises.

No.	Models of social enterprises	
	Non-profit organizations	Some of the most popular and widespread models of social enterprises are non-profit organizations that operate on the principle of self-financing through the sale of products or services. They have a built-in social mission and strive to solve specific social problems.
	Cooperatives	The model of social enterprises is cooperatives, which are owned and managed by employees. In such organizations, decisions are made democratically and profits are shared equally among cooperative members.
	Limited liability companies	Limited liability companies are a very important model of social enterprises. They pursue both business and social goals.

Source: Own study.

It should be noted that the principles of social enterprises can also be implemented by commercial companies. This depends on the philosophy of business creation adopted by managers and entrepreneurs. If they adhere to and recognize social principles in their business, they will disseminate them throughout the organization, thus generating social values in the economic space. It is also important to be aware that in today's knowledge-based economy, corporations are critically besieged by growing hegemony and expectations from better informed and diverse stakeholders, and receive more detailed emphasis on solving social problems amid internal calls for wealth generation and/or greater financial security (Awa et al. 2024). Organizational social performance models should therefore be configured with a complete business decision-making system covering all levels and functions of the company.

1.7 Creating shared value (CSV) and consumer-driven corporate responsibility (CDCR) and the perspective of high-performance companies

Some of the most important continuous and systematic activities of an organization should be the creation of values with different relations and effects. Only wisely managed organizations can see a mutual benefit, often of a long-term nature, in this perception of business. Those organizations that are strategically short-sighted in this area may suffer far-reaching consequences, including the collapse of the organization. Organizations must therefore be fully active in terms of CSR, given that their programs and public disclosures are performance-related (Jones 1995, pp. 404–437). The skillful and fair redistribution of value to all stakeholder groups is a crucial skill of mature business managers. This redistribution and allocation of value determines the continuity of the organization's functioning and its creation of long-term goals as part of the adopted strategy to a very large extent. Then, the path is generated towards the company achieving a dynamic strategic balance, which is a tool for building the value of stakeholders centered around the organization. The greater the value of stakeholders in the organization's environment, the greater the internal strength of the organization. The appropriate hierarchy of stakeholders that determines the strength of the influence of a given group of stakeholders on the growth and development of the organization is also crucial here. This is particularly important, including in relation to the place and role of customers of a given organization in the relationship with other stakeholders of the organization. This is related to the precisely designated responsibility of the organization, controlled and navigated by the expectations and requirements of customers. A sustainable lifestyle is becoming the lifestyle of choice of many consumers, especially millennials and post-millennials. This trajectory of development also means that the role of identity and consumption culture as central principles of lifestyle was highlighted by researchers as a factor to be taken into account (Gatersleben et al. 2019, pp. 37–50). The result of such responsibility is also the appropriate allocation and redistribution of the value created for all involved stakeholders of the organization in accordance with the incentives applied and the contribution received to the increase in the value of the company. Contributions and incentives are the primary causal link in determining the responsibilities of different stakeholder groups of a consciously managed organization. Therefore, from the perspective of creating appropriate rules for supporting climate action and shaping positive behavior towards climate change, the concept of building climate-smart value is very important. Climate-smart value (CSV) includes productivity, resilience and climate change mitigation (Rosenstock et al. 2020, pp. 76–81). This is related to the systemic fight against climate fragility reduction.

1.8 Conclusions

Mechanisms for shaping corporate social responsibility are some of the core areas of effective and efficient value-based management. The chapter addresses the most important dimension of corporate social responsibility in fulfilling its functions in business, society and the economy. A core area of the theoretical discussion and practical solutions is the transfer of the principles of corporate responsibility to all core components of modern management and the search for measurable benefits for companies in the space of the short-term and long-term survival of the organization on the market. It is important to place responsibility in such a way that everyone feels like winners and creators of generated values with their adequate redistribution to all groups of stakeholders and actors of business networks.

References

Austin, J., Stevenson, H. and Wei-Skillern, J.(2006). *Social and Commercial Entrepreneurship: Same, Different, or Both?* Entrepreneurship Theory and Practice 30(1): 1–22.

Awa, H. O., Etim, W. and Ogbonda, E. (2024). *Stakeholders, stakeholder theory and Corporate Social Responsibility (CSR),* Corporate Soc Responsibility, https://doi.org/10.1186/s40991-024-00094-y.

Bowen, H.R. (1953). *Social Responsibilities of the Businessman*. New York, NY: Harper.

Carroll, A.B. (1991). *The pyramid of Corporate Social Responsibility: Toward the moral management of organizational stakeholders*, Business Horizons, nr 34(4): 39–48.

Clarkson, M. (1995). *A stakeholder Framework for Analyzing and Evaluating Corporate Social Performance*. Academy of Management Review, 20(1): 106.

de Waal, A.A. (2007). *The Characteristics of a high Performance Organization*. Business Strategy Series, Vol. 8 No. 3, pp. 179–185. https://doi.org/10.1108/17515630710684178.

Freeman, R.E. (1984). *Strategic Management: A Stakeholder Perspective*, Engelwood Cliffs, NJ, Prentice Hall.

Friedman, A.L. and Miles S. (2002). *Developing Stakeholders Theory*. Journal of Management Studies, vol. 39, no. 1.

Gatersleben, B., Murtagh, N., Cherry, M. and Watkins, M. (2019). *Moral, wasteful, frugal, or thrifty? Identifying consumer identities to understand and manage pro-environmental behavior.* Environment and Behavior 51(1): 24–49.

Gatersleben, B., White, E., Abrahamse, W., Jackson, T. and Uzzell, D. (2012). *Values and sustainable lifestyles. In Transforming markets in the built environment*. Oxon: Routledge, pp. 37–50.

Green Paper Promoting European framework for Corporate Social Responsibility, Commission of the European Communities, Brussels, 18 July 2001, http://ec.europa.eu/employment_social/soc-dial/csr/greenpaper_pl.pdf.

Iborra, M. and Riera, M. (2022). *Corporate social irresponsibility: What we know and what we need to know*. CorpSocResponsibEnvironManag. pp.1–19., DOI:10.1002/csr.2428.

Jones, T. (1995). Instrumental stakeholder theory: A synthesis of ethics and economics. Academy of Management Review, 20(2): 404–437.

Latour, B. (2005). *Reassembling the Social: An Introduction to Actor-Network-Theory*. Oxford: Oxford University Press. p. 238. ISBN 9780199256044.

Manetti, G. and Toccafondi, S. (2012). *The role of stakeholders in sustainability reporting assurance. Journal of Business Ethics*, 107: 363–377.

McGrath, S. K. and Whitty, S. J. (2018). *International Journal of Managing Projects in Business* 11(3): 687–707, DOI 10.1108/IJMPB-06-2017-0058.

Miles, R.H. (1987). *Managing the Corporate Social Environment: A Grounded Theory*. Prentice Hall, Englewood Cliffs.

Mitchell, J.R., Mitchell, R.K., Hunt, R.A., Townsend, D.M. and Lee, J.H. (2022). *Stakeholder engagement, knowledge problems and ethical challenges*. Journal of Business Ethics, 175(1): 75–94. https://doi.org/10.1007/s10551-020-04550-0.

Rappaport, A. (1986). *Creating shareholder value: the new standard for business performance*. Free Press: N. Y.

Rosenstock, T.S., Lubberink, R., Gondwe, S., Manyise, T. and Dentoni, D. (2020). *Inclusive and adaptive business models for climatesmart value creation*. Current Opinion in Environmental Sustainability, 42: 76–81, https://doi.org/10.1016/j.cosust.2019.12.005.

Scott Armstrong, J. (1977). *Social Irresponsibility in Management.* Journal of Business Research, 5: 185–213. Reprinted with permission of Elsevier North-Holland Inc.

Stahel, W.R. (2006). The Performance Economy, Palgrave Macmillan, London, p. 13.

Teece, D. and Pisano, G. (1994). *The dynamic capabilities of firms: An introduction.* "Industrial and Corporate Change", No 3.

Tokoro, N. (2007). *Stakeholders and Corporate Social Responsibility (CSR): A new perspective on the structure of relationships.* Asian Business & Management, 6: 143–162. https://doi.org/10.1057/palgrave.abm.9200218.

WBCSD Stakeholder Dialogue on CSR, Holland, 6–8.09.1998, http://www.wbcsd.org/DocRoot/hbdf19Txhmk3kDxBQDWW/CSRmeeting.pdf, p. 6.

Wiklund, J. and Shepherd, D. (2003). *Entrepreneurial Orientation and Small Business Performance: A Configurational Approach.* Journal of Business Venturing, 20(1): 71–91.

Wilson, F. and Post, J.E. (2013). *Business models for people, planet (& profits): exploring the phenomena of social business, a market-based approach to social value creation. Small Bus. Econ.* 40: 715–737.

Wood, D. J. (1991). *We need to ask the right question to solve the mystery of creating the good society.* Business Horizons.

Yunus, M., Moingeon, B. and Lehmann-Ortega, L. (2010). *Building Social Business Models: Lessons from the Grameen Experience*, Long Range Planning, 43(2-3): 308–325.

2

Corporate Responsibility Seen from the Perspective of the Principles of Sustainable Development and Business in the Context of the Climate

2.1 Introduction

The aim of this chapter is to present the core principles of shaping corporate social responsibility against the background of the principles of sustainable development and business in the context of the climate. The scope of the chapter covers sustainable and business issues and their strategic perspectives for the business and economy ecosystem, with particular emphasis on the ESG concept, as well as corporate strategies in terms of dynamic balance and factors shaping the value of companies in the long term. The subject of the chapter concerns a multidimensional approach to the aspects of corporate responsibility in terms of the principles of sustainable development and business for generating business sustainability and the ability to manage business continuity in accordance with the criteria of a natural ecosystem consistent with the socio-economic ecosystem.

2.2 Sustainable development and its strategic perspectives for the business and economic ecosystem

Strategic and tactical–operational activities can be chaotic or orderly. Everything de facto depends on the internal and external context of the organization and on the ways of business management chosen by entrepreneurs or managers. Different decision-making options can be adopted. We react to external chaos with internal chaos. We react to external chaos with internal order. Rather, it is impossible to react to external order with internal order because there is no external order in the ecosystem

DOI: 10.1201/9781003474401-2

around us. The functioning of companies in a hyper-competitive market is marked by the search for new management paradigms which will show the way forward in order to achieve a sustainable competitive advantage. Contemporary directions of management derived from the theory and practice of business show the power of companies using their individual philosophies, concepts and methods in modern business. Organizations that search for optimal business models for sustainable growth and development undertake strategic challenges based on the assumptions of sustainable development and sustainable business. This leads to ensuring long-term business continuity in line with all stakeholder groups. The way of doing business is radically different today, and in the era of unrestricted access to information and knowledge, a great number of opportunities arise to verify the image and reputation of the company on the market quickly. Hence, the growing importance of the concept of corporate social responsibility, which can be used to build the long-term value of the company.

Therefore, if a strategic factor of the company's functioning related to the creation of the company's value is introduced, a clear and wide way to build the appropriate configuration and strategic architecture of the company aimed at the dynamic growth of the company's value while meeting the principles of corporate social responsibility is created.

The strategic prospects for the development of economies and societies across the globe are uncontrollable, heterogeneous and mutually inconsistent. Conscious economies such as the European Union try to make rational forward-focused decisions, taking care of the future generations of this world. Other economies, especially third world economies, are focused on the present, which often allows them to survive at all costs, due to very difficult existential conditions at both social and economic levels. In this sense, climate change has a large-scale dimension around the world. The events of recent years, environmental threats to the world and the global economic, health and COVID-19 crises, the military crisis caused by the wars in Ukraine and the Middle East, and the financial crisis are creating a new picture of the functioning of companies. The strength of external influence and the multifaceted approach to doing business highlight the ethical, moral and economic nature of the development and growth of competing companies. It should also be noted that a stream of emerging internal and external threats has led to today's organizations having to do business in the so-called multi-crisis environment. Corporate social responsibility is becoming increasingly visible and appreciated by companies in this space of business logic. This philosophy indicates a very broad dimension of the functioning of companies focused on the external environment of the company, whose entities set the directions of strategic thinking and strategic market behavior of organizations. The degree and complexity of the relationship with the external environment is reflected in the competitive strategies adopted. The opening of companies to the company's stakeholders, who may include customers, suppliers, co-operators, employees, government and non-governmental institutions and other influential institutions, and listening to their needs and expectations makes companies often start to consider functioning in business from different perspectives and points of view. They begin to notice that it is possible to build not only a competitive position but also to have a direct, significant impact on the reality of the world around us through a positive social perception measured, among others, by an acceptable reputation on the market and a positive opinion of the external environment. Moreover, companies are beginning to clearly see the important correlation between stakeholder satisfaction and the financial performance of the company, including generating social results, which in the process of operating in a difficult market full of threats translates into increased company value in the long term. This is also influenced by the value of stakeholders gathered around the company. Among the company's stakeholders, the natural environment is becoming a silent stakeholder embedded in an ecosystem where various types of organizations exist (Chodyński 2012, pp. 21–25).

However, it should be emphasized that on the Anglo-Saxon markets, in particular in the USA, the shareholder theory is promoted, which highlights the central role of company owners and the focus on pursuing their interests as part of creating company value. However, its current impact provokes the aggression of various groups of actors of the network of relationships around the organization rather than a directly positive reception and the associated benefits. The main assumption of this financial model of a corporation is the belief that shareholders, as owners, have the exclusive right to its resources and profits. Such an interpretation is often received unfavorably at this point in time. This determines the company's goal, which is only to serve mainly the interests of shareholders by maximizing the value of corporations and thus generating profits for their shareholders–owners. In the case of the financial model of a corporation, it is treated as the property of shareholders and thus seen as an instrument intended to serve the owners (Zalega 2003, p. 27).

Numerous companies that rely on corporate social responsibility supported by the international community implement the assumptions of building a competitive position, taking into account the wishes and expectations of all key stakeholders, paying attention to the importance of ecological, and now increasingly climate-related, issues. The concept of sustainable development is the starting point and platform for the implementation of such a view of building a position and competitive advantage. It should clearly translate, inter alia, into the greening of business models, strategies, processes and projects at the company level. A great number of definitions and interpretations of sustainability occur in the relevant literature. Most of them pay attention to various aspects of the application, use and understanding of these words. The concept of sustainable development was first publicly used at the first UN Conference on the Human Environment in 1972 in Stockholm. The first definition appeared at the Session of the Governing Council of the United Nations Environment Program (UNEP) in 1975 and was as follows: sustainable development is "a desirable course of economic development that does not materially and irreversibly affect the human environment, does not harm the biosphere and does not lead to the disruption of the laws of nature, economy and culture" (Poskrobko 1998, p. 75). This definition is a premise for companies to build responsible businesses that take into account ecological criteria in their activities. In 1987, the World Climate Change Commission (WCCC) was established. The UN Environment and Development Report "Our Common Future" identifies sustainable development as economic and social development that will ensure that the needs of modern society are satisfied without compromising the ability to fulfill the needs of future generations (Poskrobko 1998).

However, the Second United Nations Conference on Environment and Development in Rio de Janeiro in 1992, which developed the so-called Rio Declaration, was the most important in terms of promoting such behavior. The Rio Declaration consists of 27 principles defining the rights and duties of nations. One such principle was that present and future generations have the right to a healthy and creative life in harmony with nature (Keating 1994, p. 13). The Rio Conference:

- documented strategic aspects of sustainable development,
- emphasized its core feature, i.e., durability,
- excluded, or at least limited, the possibility of shifting the negative consequences of development to future generations (it should be noted that the widespread shift of adverse effects is identified and limited to ecological effects, while the problem is much wider – the adverse effects can be shifted to future generations, for example as a result of unjustified borrowing by contemporaries),
- places the environment and development on an equal footing (Piontek 2022, p. 29).

Sustainability has three main dimensions, namely environmental, economic and social. The environmental aspect emphasizes the protection of natural resources and ecosystems. The economic dimension focuses on generating profit, minimizing costs, encouraging research and development and driving economic growth. The social aspect is aimed at improving living standards, providing education and ensuring equal opportunities for all (Brown et al. 1987, pp. 13–719).

Currently, the most important international document on climate change with a strong link to the principles of sustainable development is the Climate Change Strategy—the Paris Agreement. The Paris Agreement is an international treaty that legally obliges signatories to combat climate change. In 2015, for the first time in history, governments jointly agreed to make a major effort to limit global warming and counteract its effects. The Paris Agreement entered into force on 4 November 2016, when the condition of ratification by at least 55 countries responsible for at least 55% of global greenhouse gas emissions was fulfilled. This is an action plan aimed at limiting global warming (https://www.consilium.europa.eu/en/policies/paris-agreement-climate/), (https://eur-lex.europa.eu/legal-content/EN/TXT/PDF/?uri=CELEX:22016A1019(01)).

Sustainable development is therefore characterized by three characteristics, namely sustainability, durability and self-sustainability of development. Sustainability means a need to maintain the right proportions, the structure of development and a balance between development needs and the need to protect the environment. The feature of durability requires maintaining the availability of environmental resources, and self-sustainability of development draws attention to the interdependence of economic, environmental and social factors in stimulating long-term economic growth and the role of ecological determinants of this growth (Adamczyk and Nitkiewicz 2007, p. 27).

According to the UN, the following are extremely important to creating a sustainable model for the functioning of nations:

- Economic growth and equitable distribution of benefits,

- Protection of natural resources and the environment,
- Social development (Johannesburg Summit 2002).

The implementation of these principles in practical terms was manifested by the development of the EN – ISO 14001 Environmental Management Systems – Requirements standard, which became a guide for companies in the implementation of pro-ecological behavior. This is where the concept of a silent stakeholder arose, i.e., components of flora and fauna in the name and on behalf of which these systems are built and developed. In the face of these dynamic changes, the concept of sustainable development has increasingly been replaced by the concept of eco-development. Eco-development must take into account the real links between society, the economy and the natural environment (Poskrobko 1997). Eco-development can be treated as a new philosophy of global, regional and local development opposing narrowly understood economic growth (Boris 1998, p. 11). However, attention is paid to the fact that it is clear that development should be reasonably sustainable financially, economically, ecologically, socially, institutionally, and politically as it would be impossible for development to be completely sustainable and such an expectation would be wishful thinking. However, prosperity, like the category of happiness, also comprises cultural and spiritual aspects, touching upon the mental condition, the emotional sphere and well-being. A number of these aspects lie outside the conventional field of observation of traditional economics, but the point is that it is worth going beyond conventions and traditions because it is the most intellectually popular (Kołodko 2008, pp. 270–271). How a company responds to the conditions and opportunities arising from the implementation of the concept of sustainable development depends on how they have been integrated into the economic policy of the state and on whether and how companies are assessed in terms of the adaptation of their activities to the requirements of sustainable development (Kuciński 2009, p. 18). The economy, currently heavily regulated at the state level, significantly affects the strategic behavior of entrepreneurs. The definition contained in the Environment Protection Law is based not only on the principle of integrating natural, economic, social and political spheres, but also on the basis of equal opportunities for access to resources and intergenerational justice (Jeżowski 2002, p. 2022). The concept of eco-development is treated here as being on an equal footing with the concept of sustainable development. This interpretation is assessed differently. So how do companies apply and integrate sustainability principles in building effective competitive strategies for their long-term growth and development? In the case of sustainable development, companies can identify specific, unique, and green instruments of competition. These include:

- the ecological quality of products and technologies,
- eco-innovation, meaning the ability of an organization to implement innovations in connection with the market in the context of external ecological trends and internal conditions of the company (Adamczyk 2001, p. 277),
- flexibility in adapting ecological products to the customer's needs in connection with activities in the area of the marketing mix,

- creating a pro-ecological image of the company in connection with a positively perceived ecological product brand,
- the reduction of operating costs through pro-ecological management.

Eco-development is considered in terms of the search for sources of obtaining a sustainable, safe position on the market based on social responsibility towards the market environment. P. Kotler cites and describes numerous examples of companies which based their competitive advantage on a pro-social and pro-ecological image on the market (Kotler 1998, p. 22). The positive impact of this image translates into increased turnover caused by a change in customer preferences. Kotler believes that the pro-social character of the market image is at the same time a stable form of preventing the copying of products. The pro-ecological and pro-social trend in practices aimed at gaining supporters occurs on both European and American markets (Hawken 1993, p. 12). The logical and methodological structure of the perception of pro-ecological and pro-social principles should be multi-layered and universal, and should be developed even in the least conscious and vulnerable regions of the world.

2.3 Sustainable business from the perspective of the business and economy ecosystem

Sustainability issues are constantly being extended to the various components of human life and activity on earth. This is particularly important in the case of the instability of almost all elements of the functioning of the world. Therefore, this concept has also become firmly rooted in the economy and business. Sustainability is the ability of an company to continuously learn, adapt and develop, revitalize, reconstruct, and reorient to maintain a sustainable and distinctive position on the market by offering above-average value to buyers today and in the future (in line with the paradigm of innovative growth) thanks to organic variability constituting business models, resulting from the creation of new opportunities and goals and responses to them, while balancing the interests of various groups (Grudzewski et al. 2010, p. 27). The concept of sustainability comes from a broader view and approach to this issue, in relation to balancing the influence of individual political, social and scientific groups over time (Kidd 1992, pp. 1–26). This means that a close correlation between business sustainability and sustainable stakeholder relationship management occurs. The core aspect of sustainability occurs when the expectations and ideas of a company related to the market and society influence prevailing opinions on what can and cannot be done in the practice of sustainable business. On the other hand, stakeholders and their expectations contribute to the answer to this question (Svenson et al. 2010, p. 338). The philosophy of the sustainability business currently plays an increasingly important role in the responsible conduct of the market struggle, which is based on 10 core principles that the company should follow:

1. Positive impact of the company.
2. Positively perceived brand and reputation.
3. Ecological processes consistent with the planned ecological effect.

4. Achieving acceptable financial results.
5. Implementing multidimensional measurements.
6. Implementing an effective and efficient competitive strategy.
7. A clear answer to the question: Can a responsible company function better or can a well-functioning company be more responsible?
8. Testing effective business scenarios.
9. Work and growth.
10. Finding and closing gaps in the area of sustainable development of the company based on indicator analyses (Burchell 2008, pp. 111–118).

Relationships and proper dialogue with stakeholders determine the shape and nature of the principles of sustainability in business. In this approach, it is also important to determine and measure the value of stakeholders, i.e., the amount of discounted cash flows appropriated from the market through relationships with stakeholders. Therefore, what is the relationship between company sustainability and stakeholder value? The higher the value of stakeholders gathered around the company, the greater the probability of the company's survival on the market. The strength of the inter-organizational bond and the continuous interaction, interdependence and long-term relationships with stakeholders contribute to the achievement of better financial and non-financial results of the company, affecting the ability to build and increase the long-term value of the company. The dynamic balancing of strategic and tactical–operational activities in the adopted business model is a difficult task and requires, above all, an understanding of the context of conducting such activities. Is this even reasonable and feasible? After all, in the era of chaos and market changes, imbalance prevails, especially when we exist in a permanent stream of various types of crises. A profitable business model can be achieved through actions aimed at meeting the principles of sustainability. This model can be called a sustainable business model, understood as a mapping in a given place, time and business space of the structure of relationships among factors ensuring that the current, internal and external needs of stakeholder groups are satisfied, which enables the company to achieve a competitive advantage and constitutes the creation of a future platform for a company's growth and development, ensuring business continuity. Such an understanding of the business model allows for a change in strategic thinking consisting in building a competitive advantage through the agile and flexible implementation of business decisions in the system from the current business model today to the current business model tomorrow (Jabłoński 2008, p. 19). This model can be implemented in business practice using a balanced strategy, understood as a bundle of interdependent strategic goals aimed at fulfilling the needs of individual groups of stakeholders related to other company goals focused on development, processes and financial results. The result of this strategy is to achieve sustainable company results that guarantee the long-term value of the company. Achieving business sustainability is about a coherent and transparent strategy for maximizing the value brought by a company operating for a long time, while optimizing the results achieved in the short and medium term. It consists of numerous factors, including the pursuit of vigilance, endurance and resilience. This strategy requires the preservation of well-maintained, stable assets, the continuous introduction of innovative, unique products and services and a good reputation among customers, employees, distributors, suppliers, authorities and other stakeholders investing in the company (Kotler and Caslione 2009).

This model increases the efficiency of the company's operations related to the determination of:

- the place and role of the different media (border conditions and processes) influencing the balancing processes,
- cause-and-effect relationships between various activities in the company,
- the impact of these actions on achieving sustainable results,
- the likely reactions of the company to stakeholders,
- the potential and actual impact of these activities on financial results (Epstein and Buhovac 2010, p. 307).

Financial results are a result of and reflect previous strategic decisions, so when these decisions are made systematically in relation to many important elements of the company's operation and dynamically affect its environment, a chance occurs to achieve and sustainably maintain an advantage over competitors. This task requires building comprehensive competitive potential visible to customers. Strategy is not what a company does, but what customers and other stakeholders see it doing. These decisions can be facilitated by remodeling the organizational structure into a sustainable organizational structure integrated into a tight, sustainable, and at the same time dynamic strategic management system. Designing an effective and efficient strategic management system aimed at increasing the long-term value of the company is one of the priority activities of a modernly managed company. When considering sustainability, it can be concluded that in order for the company's value to be sustainable, the strategic management system should also be sustainable. This requires the company to fully integrate core strategic factors that constitute the business model's turn towards sustainability in the economic, ecological and social areas:

- economic sustainability – increasing the profitability of the company through the efficient use of resources (human, raw materials, finance), effective projects and undertakings, good management, planning and control,
- environmental sustainability – avoiding harmful and irreversible consequences for the environment through the efficient use of natural resources, promoting renewable resources, protecting soil, water and skillful waste management,
- social sustainability – responding to the needs of society, including all other stakeholders (Abidin and Pasquire 2007, p. 277).

This means that it is important to combine these three areas of sustainability into one coherent strategic management system based on value-creating activities. Enhanced awareness and sensitivity in the field of sustainability affect the growth of tangible and intangible values of the company through actions. Focusing the strategic management system on assessing business efficiency by controlling value drivers such as sales growth, profit growth, increased fixed and current assets, sufficient liquidity to settle tax liabilities, cost of capital and increasing the duration of competitive advantage will undoubtedly guarantee the effect of the business sustainability of the company. When examining the sustainability business model in the long term, it is impossible to ignore the concept of corporate social responsibility as a source of creating sustainable value in the company. This concept guarantees the operation of an ethical, ecological and economic business based on values shared by stakeholders grouped around the company. The environmental dimension at the technological level is related to the principles of industrial ecology (Lifset and Graedel 2001, pp. 3–15). Then, an important component of this approach is the aggregation of the organization's personnel to build the well-being of the organization's employees. This allows a business model to be built in the network of internal and external relationships that determine the sustainable competitive advantage of the company. This creates the ability to achieve sustainable organizational performance. In this sense, the relationship between the stakeholders around the company and the responsibility of the company towards them, as well as the strategic assumptions of the company and its economic, social and environmental performance, are important. The source of the increased value of the company can be not only financial assets, but also environmental behavior at the level of the organization. An awareness of the relationship between the profitability of the company's activity and corporate social responsibility makes the rules of operation of companies on the market change (Jabłoński 2010, p. 42). This change results in a high concentration of managers' decisions on satisfying the needs of the company's shareholders, taking into account the needs of all stakeholder groups. Such strategic thinking is long-term thinking with a clear focus on the emerging future. Then, this market success becomes the success of the entire network of relationships between the company and the environment.

2.4 The ESG concept in responsible business models

The components of the natural environment, social aspects and properly understood and applied corporate governance create a contemporaryicture of understanding corporate responsibility. The area of systemic regulation in this area is of strategic importance on the scale of global understanding of the impact, especially of international corporations with a global reach. In addition, previous global economic crises, which led to numerous extremely serious problems in the world, also revealed highly pathological economic phenomena, the effects of which are often felt to this day. This needs to be addressed and responded to proactively and systematically. The literature on climate change reporting draws on points of view such as legitimacy theory, stakeholder theory and institutional theory to explain organizational practices on greenhouse gas reporting. Companies can report greenhouse gas emissions to meet societal expectations and retain legitimacy (Hrasky 2011, pp. 174–198). The requirements of the ESG concept are also defined in this scientific approach.

These requirements are regulatory and systemic at the level of established formal and legal requirements. ESG factors are the company's business standards, focusing on three main areas, namely environmental (how the company acts as a nature manager), social (relationships with employees, suppliers, customers and communities) and corporate governance (leadership, executive salaries, audits, internal controls and shareholder rights) (Wamane 2023).

ESG (Environmental, Social and Governance) is also a concept covering three core areas that support organizations in conducting their business in a responsible and sustainable way (Mousavizadeh 2024).

The first area is environmental, which analyzes the impact of activities on the surrounding nature. For example, in the context of the organization of health services, this means, among others, managing energy consumption, reducing carbon emissions and the responsible management of medical waste. Another area is social, which focuses on relationships with patients, employees and the local community. The priority here is to ensure safe and decent working conditions, support the mental health of staff and promote diversity and inclusion. The third core area is governance, which refers to the management of the organization, transparency of decision-making and adherence to ethical principles.

The dynamics of governance in global management can be considered from several analytical perspectives of power. Barnett and Duvall (2005) identified four types of governance:

- compulsory (A can exercise power directly over B with the resources of A),
- institutional (A can exercise indirect control over B through institutional arrangements such as decision-making rules, formalized lines of responsibility, divisions of labor and dependency structures),
- structural (power is determined by the social abilities and interests of the actors),
- productive (power means the production of entities through dispersed social relations, i.e., A can exercise control over B through semantic practices) (Barnett and Duvall 2005).

In hospitals, for example, this is manifested through responsible resource management, clearly defined decision-making structures and the implementation of the highest ethical standards. ESG reporting makes it possible to systematically monitor and improve working conditions, both physically, in terms of safety and ergonomics, and mentally, providing support for the mental health of employees. Increased transparency and public trust, the sustainability reporting obligation under the CSRD, allows organizations to enhance their credibility in the eyes of customers, local communities and investors. Transparency in management, environmental reporting, and community initiatives helps build trust, which is a core component of any organization's operations. ESG-compliant measures, in addition to reducing negative environmental impacts, lead to significant savings. Improved efficiency in the management of energy, waste or water resources allows for the reduction of operating costs. This allows different types of entities to focus on their main mission – to ensure the best possible value of services and products for their customers. Younger generations of employees increasingly prefer employers who act responsibly, care about the environment and offer a healthy working environment. Meeting ESG requirements and complying with the CSRD can open up new funding opportunities. The European Union and other institutions prefer to support organizations operating in accordance with the principles of sustainable development. For businesses, this means faster access to funds for modernization, investment in innovative technologies and the development of products and services offered. The EU taxonomy focuses on the concept of ESG in a broad sense. This is a colloquial name for the regulation on the establishment of a framework to facilitate sustainable investment. The main objective of this regulation is to support investments in projects and activities that contribute to the EU's climate and environmental objectives. The key here is to build nature-based solutions, returning to the natural ecosystem in many cases. Nature-based solutions are activities that use natural processes and ecosystems to solve environmental, social and economic problems. The European Commission defines nature-based solutions as nature-inspired and nature-supported solutions that are cost-effective while delivering environmental, social and economic benefits and helping build resilience. Such adopted solutions bring ever more diverse natural features and processes to cities, landscapes and marine landscapes through locally adapted, resource-efficient and systemic interventions. Nature-based solutions must benefit biodiversity and support the provision of a range of ecosystem services. Such solutions are used to build the individual climate adaptation plans of cities and regions. Table 4 presents list of core legal provisions related to the ESG concept in the European Union.

Table 4. List of core legal provisions related to the ESG concept in the European Union.

No.	Legal provision related to the ESG concept
1.	Directive 2022/2464 of the European Parliament and of the Council (EU) of 14 December 2022 amending Regulation (EU) No 537/2014, Directive 2004/109/EC, Directive 2006/43/EC and Directive 2013/34/EU as regards corporate sustainability reporting. Available at: https://eurlex.europa.eu/eli/dir/2022/2464/oj?locale=pl.\
2.	Directive of the European Parliament and of the Council (EU) adopted on 24 April 2024 on corporate sustainability due diligence and amending Directive (EU) 2019/1937 (Corporate Sustainability Due Diligence Directive – CSDDD). Available at: https://www.europarl.europa.eu/RegData/seance_pleniere/textes_adoptes/definitif/2024/04-24/0329/P9_TA(2024)0329_EN.pdf
3.	Regulation 2022/1288 of the European Parliament and of the Council (EU) of 6 April 2022 supplementing Regulation 2019/2088 of the European Parliament and of the Council (EU) with regard to regulatory technical standards specifying the details of the content and presentation of the information with regard to the "do no significant harm" principle. Available at: https://eur-lex.europa.eu/legal-content/PL/TXT/?uri=CELEX%3A32022R1288.
4.	Regulation 2019/2088 of the European Parliament and of the Council (EU) of 27 November 2019 on sustainability-related disclosures in the financial services sector. Available at: https://eur-lex.europa.eu/legal-content/PL/TXT/?uri=CELEX%3A32019R2088.
5.	Regulation 2020/852 of the European Parliament and of the Council (EU) of 18 June 2020 on the establishment of a framework to facilitate sustainable investment, and amending Regulation (EU) 2019/2088. Available at: https://eur-lex.europa.eu/legal-content/PL/TXT/?uri=celex%3A32020R0852.
6.	Commission Delegated Regulation (EU) 2021/2139 of 4 June 2021 supplementing Regulation 2020/852 of the European Parliament and of the Council (EU) by establishing the technical screening criteria for determining the conditions under which an economic activity qualifies as contributing substantially to climate change mitigation or climate change adaptation. Available at: https://eur-lex.europa.eu/legal-content/PL/TXT/?uri=CELEX%3A32021R2139.
7.	European Sustainability Reporting Standards (ESRS). Full study available on the Ministry's website. Available at: https://www.gov.pl/attachment/311a77d1-365c-4575-be40-ffe2ff1bea64.

Source: Own study.

It should be remembered that the concept of smart and interactive regions and cities, which become so-called hubs operating on the basis of principles determined by the network paradigm that favors the implementation of the ESG concept, applies here. The ESG concept should also be looked at in financial terms (Cerrato Garcia 2024). ESG indices are an important indicator of sustainable finance. With the increasing importance of sustainable finance in global financial markets, the need to market "green" products that incorporate ESG factors into their design is increasing, and therefore stock indices based on ESG criteria are an important part of the overall and comprehensive transition process towards a greener and more sustainable economy. One of the most important, crucial documents regulating this area in the financial market is the Non-Financial Reporting Directive (NFRD) (Directive 2014/95/EU), which imposes obligations on entities related to reporting non-financial data. This is how "green democracy" is created (Ituarte-Lima and Mares 2024).

Since January 2024, the effective and efficient management of companies in the European Union requires comprehensive compliance with the requirements of two core legal acts. Sustainability reporting requirements will apply to various business entities, and reporting organizations will be required to apply the new rules and requirements from 2025 for operations in 2024.

This is related to the two most important directives that companies must take into account in their management processes from 2024, namely:

1. CSDDD (Corporate Sustainability Due Diligence Directive) – the directive introduces verification obligations for companies that concern respect for human rights and the environment throughout the supply chain. It applies to all large companies, including listed companies, which are obliged to report ESG issues (Environmental, Social, Governance).
2. CSRD (Corporate Sustainability Reporting Directive) – introduces detailed ESG reporting requirements. This Directive aims to increase the transparency and reliability of data, enabling informed decisions by investors and consumers. From 2024, this reporting under the CSRD is a mandatory requirement for large companies and public entities, and their reports must be verified by independent experts. In order to implement the requirements contained in these Directives effectively and efficiently, it is necessary to develop two road maps, namely:
   - a road map for the comprehensive fulfilment of CSRD and CSDDD requirements,
   - a roadmap for the construction and implementation of comprehensive measurement and monitoring of CSRD and CSDDD requirements.

These directives, together with the accompanying additional legal acts, not only strongly shape but even enforce the regulated and legally controlled dimension of the business models and strategies used by companies.

Data reporting in the field of sustainable development according to the requirements of Directive (EU) 2022/2464, the so-called CSRD, covers the implementation of the project in three pillars and six modules.

The Figure 2 shows the three pillars of data aggregation necessary for correct and comprehensive sustainability reporting.

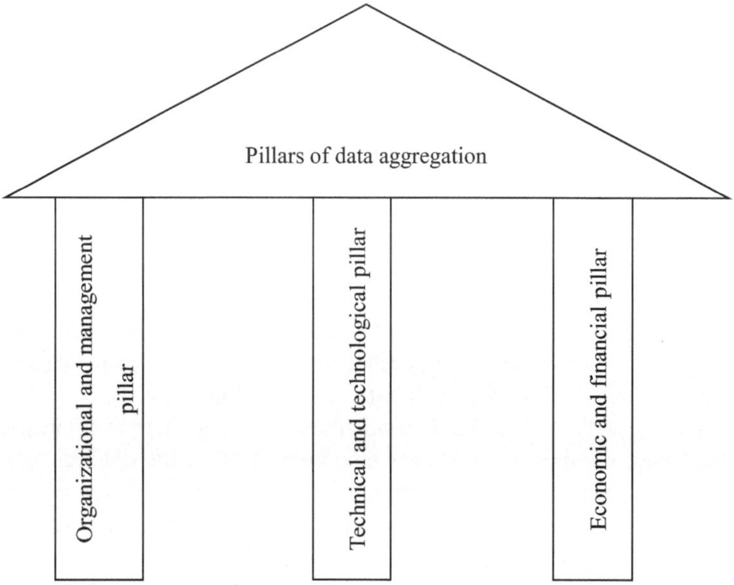

Figure 2. Pillars of sustainability data reporting. Source: Own study.

The Figure 3 shows the six modules needed to produce a sustainability report. These three defined core pillars of data aggregation in ESG reporting should have a clear focus on ESG components, namely climate change mitigation, climate change adaptation, water and marine resources, resource use and the circular economy, pollution, biodiversity and ecosystems.

Figure 3. Sustainability report modules. Source: Own study.

These modules form a mutual causal relationship related to the identification of legal provisions for data collection, the development of a data collection plan, the coordination of the data collection process, the verification of the collected data, the inference of the collected data and finally the development of a sustainability report.

These data should be clearly focused on reporting on one's own operational activities and reporting on the value chain, including products, services, business relationships and supply chains. This reporting must take into account the so-called double materiality assessment, i.e., it should include information necessary to understand the company's impact on sustainability issues and how sustainability issues affect the company's development, performance and position. Thus, we are dealing here with the mechanisms of influence, risk and opportunity. It should be noted that ESG non-financial reporting requires the assurance of an ESG report by an auditor with regard to the correctness of conducting a double materiality assessment in accordance with the European Sustainability Reporting Standards (ESRS), verifying the compliance of sustainability report disclosures with ESRS requirements, assessing the compliance of disclosures with the EU Taxonomy and finally checking the correctness of the report tagging.

2.5 Corporate strategies, from sustainable development to CSR and ESG to corporate citizenship and corporate netizenship

The strategic thinking of company managers determines the logic of long-term persistence in the market advantage of many companies. In order to survive, you have to earn money. To win, you have to consistently succeed by offering an attractive value proposition to demanding customers in harmony with local and global communities. Therefore, it is worth building the potential of market competition on stable pillars strengthened by sustainable, dynamic business decisions. The concept of sustainable development is problematic and often interpreted very differently, because development is often non-linear and generated by leaps and bounds. Therefore, the element justifying its functioning and recognition in scientific cognition is its dynamics, which may be revolutionary and/or evolutionary. These dynamics materialize in an increasing commitment to their understanding and application by new, different groups of communities around the world. Corporate social responsibility and the concept of sustainable development and business have had a lasting impact on the adopted management mechanisms resulting from the adopted paradigms. The spread of this concept to the various fields of human existence has led to the recognition of these principles by corporate communities in both the real and virtual spaces.

Corporate citizenship in this case emerges as a guiding force in the way a corporation presents itself to the world. Corporations focused on creating modern business exist in a virtual environment and then create a structure based on virtual corporate citizenship. Real and virtual corporate citizenship shapes new, sometimes surprising consumer and social behavior de facto in many aspects of life at the levels of both business and society. In the case of consistency of mutual expectations, values and responsibilities, these actions have a symbiotic and synergistic effect. When there is no such consistency, discrepancies can become a source of various types of conflicts. The economic problems associated with the intensification of various types of crises, the highly dynamic growth of the impoverishment of society and the lack of access to the basic resources of human existence create social tensions in many cases and provoke dissatisfaction with decision-makers at various levels of management. Therefore, the opening of companies to social, ecological and climatic factors while maintaining the ability to generate so-called sustainable profit in economic terms is a very important art of the appropriate, fair and transparent redistribution of revenues. It should also be remembered that the profit generated by the company is residual (final), i.e., it can be paid only if all financial liabilities towards all other stakeholder groups around the company are covered in advance.

2.6 Sustainable development and the circular economy as the sources of climate innovation based on circular business models

The linear approach to organizational management does not entirely fulfill its role at the level of processes and projects in various types of organizations. This applies to the technological, organizational and management dimensions, and is also related to the economic and financial approach. The circular economy is an approach to resource management and production that puts sustainability first. In the context of growing problems related to the overconsumption of raw materials and the production of waste, the circular economy model is gaining in importance, offering an alternative to the traditional linear model. The transition from the linear to the circular model is very important for organizational–management, technological and economic–financial changes at the levels of both states and companies. The circular economy (CE) can be a driver of sustainability, and CE can be promoted and supported by creating new and innovative business models that embed CE principles in their value propositions throughout the macroeconomic value chain (Manninen 2018, pp. 413–422), as well as the organizational value chain. The circular economy should be examined in terms of the system and process because it also refers to other approaches and concepts, namely to the concepts of clean production/the clean economy, low emission economy, the bio-economy and sustainable development. This systematism is used to achieve goals aimed at the effective use of resources and processivity to change the structural mapping of processes in the organization. Systematism should take into account the multidimensional analysis and synthesis of the entire life cycle of the organization and its business model. Life Cycle Assessment is "a tool for the analysis of the environmental burden of products at all stages in their life cycle—from the extraction of resources, through the production of materials, product parts and the product itself, and the use of the product to the management after it is discarded, either by reuse, recycling or final disposal (in effect, therefore, 'from the cradle to the grave')" (Guinée 2002, pp. 311–313). Thanks to a systemic and process-based approach, the circular economy and its practical solutions fully integrate social, economic and natural aspects at every level of management in an organization focused on responding to climate and ecological changes:

- the micro level – by creating eco-designs, eco-products, minimizing waste, introducing an environmental management system, etc.
- the meso level – through the creation of industrial eco-parks;
- the macro level – the creation of eco-cities, eco-municipalities and eco-regions (Geng and Doberstein 2008, pp. 231–239).

This is a spiral-loop economy that minimizes the flow of materials, energy and environmental degradation without limiting economic growth or technical and social progress (Lieder and Rashid 2015, pp. 36–51).

The circular economy aims to decouple prosperity from resource consumption, i.e., how we can consume goods and services and yet not depend on the extraction of virgin resources and thus ensure closed loops that will prevent the eventual disposal of consumed goods in landfill sites. Production and consumption also have associated "contamination transfers" to the environment at each step. In that sense, the circular economy is a movement towards the weak sustainability described earlier. It proposes a system where reuse and recycling provide substitutes for the use of raw virgin materials. By reducing our dependency on such resources, it improves our ability, and the ability of future generations, to meet their needs. The circular economy makes sustainability more likely (Sauvé et al. 2016, p. 53).

In a circular economy, growth is decoupled from the use of scarce resources through disruptive technology and business models based on longevity, renewability, reuse, repair, upgrade, refurbishment, capacity sharing and dematerialization (Accenture Strategy). Linder and Williander define a circular business model as "a business model in which the conceptual logic for value creation is based on utilizing the economic value retained in products after use in the production of new offerings" (p. 2) (Linder and Williander 2015). Den Hollander and Bakker describe a circular business model as "how an organization creates, delivers, and captures value in a circular economic system, whereby the business rationale needs to be designed in such a way that it prevents, postpones or reverses obsolescence, minimizes leakage and favors the use of 'presources' over the use of resources in the process of creating, delivering and capturing value" (den Hollander and Bakker 2016, pp. 1–8). According to Roos (p. 257), "[a] circular value chain business model (or a green business model) is one in which all intermediary outputs that have no further use in the value creating activities of the firms are monetized in the form of either cost reductions or revenue streams" (Roos 2014, pp. 248–274). A circular business model is how a company creates, captures and delivers value with the value creation logic designed to improve resource efficiency through contributing to extending the useful life of products and parts (e.g., through long-life design, repair and remanufacturing) and closing material loops (Nußholz 2017, p. 12). Circular business models have proven they can potentially be consistent with a firm's economic performance. Indeed, they can participate in its competitiveness through cutting production costs. More ambitiously, they can generate additional turnover by penetrating new markets and gaining new customers, and protect corporations against the volatility of raw material prices in traditional markets. Some of these strategies seem to enable improvements in operating profit and profitability (Beulque and Aggeri 2016).

Laubscher and Marinelli identified six key areas for the integration of the circular economy principles with the business model:

1. Sales model – a shift from selling volumes of products towards selling services and retrieving products after first life from customers,
2. Product design/material composition – the change concerns the way products are designed and engineered to maximize the high quality reuse of a product, its components and materials,
3. IT/data management – in order to enable resource optimization, a key competence is required, namely the ability to keep track of products, components and material data,
4. Supply loops – turning towards the maximization of the recovery of one's own assets where profitable and to the maximization of the use of recycled materials/used components in order to gain additional value from product, component and material flows,
5. Strategic sourcing for own operations – building trusted partnerships and long-term relationships with suppliers and customers, including co-creation,
6. HR/incentives – a shift needs adequate culture adaptation and development of capabilities, enhanced by training programs and rewards (Laubscher and Marinelli 2014).

Defining and monitoring the operational aspect of the circular economy is essential in order to find the best ways to maximize the benefits for the industrial value chains involved, the environment and society as a whole. This idea is based on:

- the use of energy from renewable sources;
- the use of recyclable materials or materials that are readily biodegradable without having a negative impact on the environment;
- the use of materials that can be recycled at end-of-life;
- or regeneration – they gain a so-called new life and are re-applied;
- extending the service life in relation to their normal cycle;
- increasing the intensity of use by sharing individual elements used in one process with other processes;
- increasing the possibility of using waste heat from various sources;
- industrial processes and data centers;
- supporting capacity building for circular energy communities, in particular in rural areas (Schulte 2013).

Digitalization can certainly accelerate the transition towards a more sustainable and circular economy. It can help to effectively and efficiently close material loops, providing accurate information about the availability, location and condition of products. Digitalization also enables more efficient processes in companies, helps minimize waste, promotes longer product life and minimizes transaction costs.

In this way, digitalization strengthens circular business models by helping to close the loop, slow down the material loop, and narrow the loop through increased resource efficiency (Antikainen et al. 2018, pp. 45–49). Circular economy strategies can help lower-income countries "jump" towards a more sustainable development path that avoids blocking the resource-intensive economic practices of the dominant linear system of consumption and production. Since the economies of lower-income countries are in many respects still more "circular" in terms of resource management and production and consumption practices than their developed economy counterparts, the question is how to turn this into a growth opportunity (Preston and Lehn 2017).

The concept of innovation was first used in 1912 by J. Schumpeter, who treated innovation as the formation of a new function of production, calling it a "new combination" of factors of production (Schumpeter 1912). Innovation can also be defined as "forming, accepting and applying new concepts, processes, products and services. Innovation includes the ability to change and adapt" (Thompson 1969). Innovation is not only the first application of a product or manufacturing method, but also subsequent applications by other companies, industries or countries (Johnston 1966, pp. 158–176). Innovation in business models is primarily reflected in the use of digital technologies to enable the implementation of a wider range of business models than those that were previously available to the company. An important emerging trend is the increasing adoption of many business models as a portfolio within a single company. This happens in companies of all sizes when one company uses multiple business models to serve different market segments, sell different products or engage in multi-sided markets, or to apply different business models over time. The overall framework of a business model is refined and expanded through a recursive learning process that can serve as both a cognitive instrument for understanding business models and a tool for planning innovation in business models (Li 2017). Digital climate innovation is a specific group of innovations. Breakthrough innovations are needed for the economy and society, which will ensure the change and construction of almost all elements of the business model and thus its new configuration and interfaces. These new business models can provide companies with the opportunity to diversify or expand their markets to include markets that positively perceive climate change at the company level. They can also be implemented and validated in parallel with the actual business model. In this way, the company could experiment with a new business model in the space of creating climate innovation, while the old business model would still provide revenues at an acceptable level (Ariaset et al. 2016). Key areas of climate innovation (Top 10 Climate Tech Trends in 2025) and descriptions shows Table 5.

Digital technologies are revolutionizing the traditional interdependencies between businesses. As a result, managers have begun to recognize their business environments as digital ecosystems. For companies accustomed to seeing their business environments as industries, this means a significant change of perspective – one that requires an understanding of the new strategic initiatives needed to compete in the digital age (Subramaniam et al. 2018). Digital technology has opened the way to a new understanding and perception of the economy and business and has also changed a number of core paradigms in management sciences. Digital and technological management based on automation, robotics, technical diagnostics and artificial

Table 5. Key areas of climate innovation (Top 10 Climate Tech Trends in 2025) and descriptions.

No.	Key areas of climate innovation (Top 10 Climate Tech Trends in 2025)	Description
1.	Green Construction	Green construction refers to construction activities that maximize resource conservation and reduce negative environmental impacts while ensuring quality, safety and other essential requirements. Since the 1980s, much of the research on construction management has focused on green construction (Mokhlesian and Holmén 2012, pp. 761–775).
2.	Carbon Capture, Utilization and Storage	Carbon Capture, Use and Storage (CCUS) technologies have emerged as key components of efforts to reduce CO2 emissions. These technologies are designed to capture CO2 directly from emission sources such as power plants and industrial facilities, preventing its release into the atmosphere (Yaashikaa et al. 2023).
3.	Clean Energy	Clean energy can be defined as energy that comes from renewable and zero-emission sources. Energy saved through high efficiency measures also falls in the clean energy category. The current trend of using fossil fuels as the main source of energy is unsustainable and harmful to the planet, as well as to its inhabitants, and exacerbates climate change, which is one of the greatest threats to mankind, according to the United Nations Chronicle. Clean energy can be described as a pathway to mitigate climate change. It is sufficient to look at the growing number of vehicles and industries, which shows that the economy is growing, but the hidden load, i.e., global warming, is also there (Kumari).
4.	Sustainable Land & Water Management	The concept of sustainable land management can be defined as the use of land resources, such as soil, water, animals and plants, to produce goods to meet changing human needs, while ensuring the long-term production potential of these resources and maintaining their environmental functions. Sustainable land management is a cornerstone of sustainable agriculture and a strategic element of sustainable development and poverty alleviation (Burnham et al. 2003).
5.	Building Management	The building management system was introduced to the world in the 1970, initially starting with very limited functions, but over time many changes and modifications were made, ranging from power and lighting control to building heating and cooling, as well as an alarm system as a further modification to ensure maximum safety. Intrusions can be easily monitored using a closed CCTV (Building Management Systems) system.
6.	Low-Carbon Manufacturing	Low-carbon production (LCM) can be described as a process that emits low carbon intensity (CO2) from system sources and during the production process. In addition, the term LCM can be broadly understood not only in environmental terms, but also energy savings and efficient production, as the process simultaneously uses excess energy of the available capacity/limitation (low energy efficiency) without an optimal algorithm to run the process or system, which can lead to high carbon intensity in the atmosphere (Tridech and Cheng 2008).

Table 5 contd. ...

...Table 5 contd.

No.	Key areas of climate innovation (Top 10 Climate Tech Trends in 2025)	Description
7.	Climate-Smart Agriculture	Climate-friendly agriculture is based on sustainable agriculture and rural development objectives, which, if achieved, will contribute to achieving the Millennium Development Goals (MDGs) of reducing hunger and better managing the environment. More productive and resilient agriculture relies on sound management of natural resources, including land, water, soil and biodiversity. Conservation agriculture, agroforestry, improved livestock and water management, integrated management and ecosystem-based approaches to fishery and aquaculture can make an important contribution to this area (Climate-Smart Agriculture).
8.	Circular Economy	The circular economy (CE) is currently a popular concept for future sustainable development promoted by the European Union and by several national governments (China, Japan, UK, France, Canada, the Netherlands, Sweden, Finland, etc.). Also, many business enterprises around the world have introduced CE in their manufacturing sectors to save energy and raw feeding materials. But for the moment, the concept of the circular economy is a collection of vague ideas from several technological fields and semi-scientific concepts (Korhopnen et al. 2018).
9.	Sustainable Mobility	Smart Connectivity deals with trade and policy and how to streamline economic and regulatory processes, supporting economic growth and connecting with other SDGs, as well as physically connecting infrastructure in a smart way. Sustainable mobility focuses on how to move people and goods in an efficient, safe, secure, affordable and environmentally friendly way using inland waterway transport (Unece Nexus 2021).
10.	Carbon Data and Analytics	The development of big data and smart cities has developed over time. The concept of smart cities is also linked to big data as a key factor. Synergies between smart cities and big data can be used to make cities smarter, such as cities' ability to track and measure carbon emissions. Big data analytics can be used to track carbon emissions by collecting and analyzing large amounts of data related to a company's or organization's energy consumption (Ahvenniemi et al. 2018).

Source: Own study based on https://www.startus-insights.com/innovators-guide/climate-tech-trends-innovations/#:~:text=Top%2010%20Climate%20Tech%20Trends%20%26%20Innovations%20in,...%208%208.%20Carbon%20Data%20%26%20Analytics%20

intelligence solutions has generated a new space for strategic and tactical–operational thinking at the company level, creating significant breakthroughs in science and in practice. The digital transformation, together with the climate transformation, has triggered a technological revolution of unprecedented proportions, creating strong new innovation potential. There has been a wide stream of new, innovative solutions and technologies reshaping the economy, marking it with the semantics of Industry

4.0 and Industry 5.0. For this reason, there was a transition from the so-called CSR 1.0 to CSR 2.0, and technology became the platform for this transition. Wayne Visser draws attention to five differences between CSR 1.0 and CSR 2.0. These include connectedness, scalability, responsiveness, duality, and circularity (Visser 2011). CSR is dead! Long live CSR! (Visser 2010). Such a broad approach to this subject gives rise to far-reaching scientific discussion generating the application potential focused on the spectrum of climate innovation in terms of mutual responsibility.

2.7 Long-term value creation and sustainability and business models

The capital value of a company is the result of a difficult process of effective and efficient management even in the most difficult socio-economic conditions and situations. Value is determined by the ability to offer better products and services than competitors and thus become the first choice for the often sophisticated needs of a difficult customer. Satisfying the needs of even the most difficult customer is confirmation of a quality leadership strategy adopted in the right way by many wisely managed organizations. In this context, managers have begun to notice climate change as a new platform for creating innovative market competition potential. Climate change has therefore become important to shareholder profits and value as regulatory market awareness grows and the reputational and physical risks that climate change poses to business competitiveness (Lash and Wellington 2007, pp. 94–102). Therefore, shaping the company's value in accordance with the principles of sustainable development is intellectually interesting and poses a challenge to environmentally conscious managers to some extent. Company value creation is therefore not an easy task but a necessary one. Many companies have significant problems in capturing and maintaining the right sources of its construction. This is due, on the one hand, to an incomplete knowledge of sectoral conditions and an understanding of the surrounding ecosystem, which affect the appropriate management mechanisms. This concerns mainly data, information and knowledge on the maturity of the sector, the attractiveness of the sector, the competitiveness of the sector and its stability, but also the ability to act effectively and respond appropriately to existing pressures, constraints, and opportunities. One of the core directions of corporate interest in response to the climate challenge was to improve eco-efficiency as a way not only to reduce the carbon footprint of corporations, but also to improve productivity and profitability (Dauvergne and Lister 2013). An important factor in closing the current strategic and operational gap of the company in this area may be the combined application of two management concepts, namely value-based management (VBM) and corporate social management (CSR). There are a large number of different definitions of the VBM concept. VBM is a culture and mindset that permeates the entire company from top management to rank-and-file employees (Morin and Jarrell 2001). The way of thinking must be adopted by managers and communicated to the other members of the organization in such a way that all create value for the company together by implementing a total strategy of creating value for everyone interested in the functioning of the company. Value-based management is a modern management system containing tools and procedures for making strategic and operational decisions aimed at the long-term growth of the company's value and increasing the wealth of its owners (Szablewski 2000, p. 15). This system focuses on tight and comprehensive fulfillment of all business decisions that navigate the strategy to the path of creating a stream of positive cash flows. Therefore, three aggregated elements are commonly used in element-based approaches to capture a company's business model, namely the value proposition, value architecture and profit equation (Spieth and Schneider 2016, pp. 671–696).

Appropriate organizational and management mechanisms related to the concept of value are created, namely:

- value creation – the process by which a company is able to achieve a return on capital that is higher than the cost of capital. Such a positive difference, which can be retained in the company or shared among shareholders, is the value that shareholders look for when they make a decision to invest in a company. The value creation process involves the ability to offer customers a product at a price that satisfies the condition of creating such a positive difference.
- value preservation – the need to ensure that what has been produced is not wasted or lost due to inefficiencies or poor management;
- value realization – the need to ensure that investors are able to monetize value by increasing the value of their shares, the selling prices of their shares and the payment of dividends (Bachman et al. 1997, p. 76).

Value-based management is defined as the concept of managing a business by focusing management activities and processes on maximizing its value from the point of view of the interests of the owners and the capital employed by them. It assumes the control of the operational and investment activities of the company in order to achieve the basic goal of its existence—increasing value. The concept of value-based management is based on the assumption that the owners are the only party that maximizes their own benefits and maximizes the claims and expectations of other entities associated with the company (Dudycz 2001, p. 9). Consideration should be given to the deep operationalization of the value orientation towards the framework for achieving shareholder growth objectives (Velthuis and Wesner p. 95). Company value is an intangible asset, independent of the cost of its production, allocated to a company that operates based on the principle of continuation, outside which it does not exist on its own (Kamela-Sowińska 1996). The objectives expressed in the strategy constitute a common monolith with the resources assigned to them to achieve them, and the company's personnel becomes a dynamic generator of value growth, measured by the financial potential of strength against market competitors. Managers' focus on value creation is expressed in a constant, arduous search for opportunities to achieve better financial results than competitors in the short term. In the long term, it is important to increase the market value of the company ever further. Therefore, a company's tangible and intangible resources develop in a manner compatible with the development of relationships with all entities located in the company's environment and that contribute to the growth of its value. The value of a company corresponds to the idea of an income approach to valuation, which associates value with the results of its operations. According to this approach, a company is worth as much as the income it generates, starting from today to infinity

(Skoczylas 2007, p. 15). The most important factors determining the intrinsic value of a company include the strength and durability of the economic aspects of the company's operations, the quality (i.e., honesty and competence) of managers and its financial condition (Arnold 2010, p. 20). There is a close relationship between the VBM concept and the CSR concept. The CSR model, as a means of balancing the conflicting interests of stakeholders, assigns a completely different and ambiguous role to management. The manager is to deliberately – and above all arbitrarily – decide on the distribution of benefits between competing groups of stakeholders: customers, employees, suppliers, the local community and owners (Smith and Lenssen 2009, p. 274). Then, the principles of sustainable development can be implemented in a specific company, where, e.g., on behalf of the so-called silent stakeholder, i.e., the natural environment, in addition to the state, there is a manager managing the company. If the company's business model is focused on the principles of ethics, ecology and economics, and also now on the principles and issues of climate change that underlie the principles of sustainable development, the platform for the growth of the company value will be sustainable against these three factors as well as a fourth one, namely the climate. Another image arises in defining the principles of sustainable business, because in this case the concept of sustainability is treated as permanent. A sustainable business is a sustainable and profitable business that generates shareholder value, guarantees business continuity and meets the needs of other stakeholder groups. The link between sustainable development and sustainable business in the adopted business model can be illustrated in the matrix in the Table 6.

Table 6. The link between the principles of sustainable development and sustainable business in the adopted business model.

AREA OF BALANCE	VALUE-EXPECTING AND VALUE-CREATING ENTITIES	
	SHAREHOLDERS	STAKEHOLDERS
ETHICS	Ethical value	Ethical value
ECOLOGY	Ecological value	Ecological value
ECONOMY	Economic value	Economic value
CLIMATOLOGY	Climate value	Climate value

Source: Own study.

The matrix shows that value-expecting and value-creating actors can implement value in a sustainable manner. However, it should be noted that this value has a different nature and dimension. The financial dimension occurs in the area of the economic equilibrium, then economic value is created. On the other hand, the non-financial dimension is present in the area of ecological, ethical and climate balance. This creates ethical, ecological and climatic value. If we start to determine the factors of these values, the business model will be based on ethical, ecological, climate and economic criteria.

The Table 7 presents a list of examples of ethical, ecological and economic criteria that underlie the design of a sustainable business model.

Table 7. Examples of ethical, ecological, climate and economic criteria on the basis of which the principles of building sustainable business models can be determined.

CRITERIA	
ETHICAL	1. The quantity and quality of information disclosed. 2. The number and scope of meetings with stakeholders. 3. The number and type of human rights violations. 4. The number of accidents at work caused by the employer. 5. The value of overdue receivables. 6. The number of hours devoted to building relationships with stakeholders. 7. The number of social initiatives in the company. 8. The number of awards received for philanthropy. 9. The amount of compensation paid to employees. 10. The value of training for staff development. 11. Percentage of initiatives per employee related to their professional development. 12. The number of new jobs. 13. The degree of employee satisfaction. 14. The number of employees made redundant. 15. The number and type of staff complaints. 16. The number of trade union claims taken into account. 17. The degree of compliance with payment deadlines for suppliers and co-operators. 18. The number and scope of charitable organizations supported. 19. The value of court damages as a result of negative contacts with the community. 20. The number and extent of consumer complaints. 21. The value and number of complaints settled on time. 22. The results of CSR audits. 23. The number of social programs in which the company participates. 24. The quality of the customer service indicator.
ECOLOGICAL/CLIMATE	1. The effectiveness of achieving environmental and climate objectives. 2. The degree of compliance with legal regulations in the field of environmental protection and climate requirements. 3. The degree of use of pro-ecological raw materials. 4. Environmental and climate impacts in terms of air emissions. 5. Environmental and climate impacts in terms of emissions to soil. 6. Environmental and climate impacts in terms of emissions to water. 7. Environmental and climate impacts in waste management. 8. The degree of consumption of natural resources. 9. The degree of motivation of employees to achieve the set ecological and climate goals. 10. The degree of creation, planning, control and acquisition of ecological and climatic information. 11. The degree of implementation and functioning of environmental and climate standards. 12. The scope of information provided on the state of environmental and climate protection in the company to public and local authorities and external investors. 13. Causing environmental and climate risks. 14. The product life cycle (e.g., LCA). 15. The eco-friendliness of products. 16. The greening of business processes. 17. The application of Best Available Techniques (BAT). 18. Carrying out ecological and climate reviews. 19. The occurrence of environmental failures. 20. Paying for the use of the environment and climate. 21. Administrative penalties for exceeding or breaching the conditions laid down in permits or other decisions of administrative authorities. 22. The degree of ensuring recovery (mainly recycling of packaging waste). 23. The monitoring of the company's environmental nuisance with the use of nuisance indicators. 24. Civil liability for damage caused by environmental and climate impacts. 25. Administrative responsibility for adverse environmental and climate impacts. 26. The investment activity of the company in environmental and climate protection. 27. The environmental and climate credibility of the company. 28. The sensitivity of top management to environmental and climate problems. 29. The awareness of employees in the field of environmental protection and the climate. 30. The burden borne by the company in the assessment of the external environment.

Table 7 contd. ...

...Table 7 contd.

CRITERIA	
ECONOMIC	1. The effectiveness of achieving financial objectives. 2. The degree of use of the property. 3. The degree of accomplishment of sales. 4. The basic growth rate. 5. The degree of cash flows from operating activities. 6. Cash return on sales. 7. The level of dividend payment. 8. Cash return on assets. 9. Value of trade in receivables. 10. Average recovery period. 11. Stock turnover. 12. Liabilities turnover. 13. Liquidity ratio. 14. Productivity of working capital. 15. Working capital. 16. Interest coverage ratio. 17. Financial debt ratio. 18. Net assets. 19. Profitability of the assets involved. 20. Earnings per share. 21. The value of the company to profit. 22. Leverage ratio. 23. Economic added value.

Source: Own study

The criteria in the table above can be a kind of compass and company control system to determine the appropriate configuration of a defined business model in the context of maintaining ethical, ecological, climatic and economic balance. A sustainable business model should therefore take into account the principles of sustainable development and sustainable business, becoming a platform, the foundation for the pursuit of a sustainable strategy towards all groups of stakeholders, ensuring long-term value growth. The chaos and turbulence of the external environment define the dimension of difficult competition of modern companies. The search for a path to success measured by the increase in value is hindered by the multidimensional image of the market space in which there is a network of stakeholders, including hostile stakeholders. Companies therefore need to build their business to meet their specific needs, while ensuring the sustainability, profitability and efficiency of this model. This model, which meets the principles of dynamic equilibrium, becomes a crucial factor in maintaining business continuity. The use of the principles of sustainable development and sustainable business makes this model a universal model aimed at creating long-term value. The combined and continuous pursuit of the potential inherent in these concepts can be an important factor in companies on the market achieving a sustainable growth rate. From the perspective of key limitations, however, the result of such an approach is unfortunately the promotion of the so-called double technocapitalism, including the associated political concepts, in many circles. This concept strongly promotes the concept of capitalism based on the technocratic management of states, regions, cities and companies supported by the very instrumental use of technology mainly to increase the wealth of shareholders of companies on a global scale.

One of the most important dimensions of climate change requiring managerial interventions in practical terms is the development of an appropriate system for measuring and monitoring key indicators of climate effectiveness and efficiency. It is manifested in the set and achieved target values of individual indicators expressed in absolute and relative values, as a percentage, in units of measure and in monetary value. Individual indicators can be expressed in Key Process Indicators (KPI). They should also be correlated with Key Risk Indicators (KRI). Then, their proper level of feasibility is determined. The designated measurement system should be tight, reliable and transparent, as it is the basis for decisions that, in the case of climate and its changes, may have a direct or indirect impact on human life or health.

2.8 Conclusions

Issues and core concepts in the area of sustainable development and business play a very important role in perceiving business and the economy in a wise and mature interpretation, without deliberate populism and overinterpretation. The shape of modern business should result directly from the logical perception of the entire network of internal and external relations of the organization. Then, the appropriate strategic vision of the development and growth prospects of the organization is directed and embedded in the exchange of mutual and positive values and is determined by the future of the world and not only the present existence of business on earth. This strategic vision is then built into the decision-making processes of different groups of decision-makers at different levels and functions of the organization.

References

Abidin, N.Z. and Pasquire, Ch.L. (2007). *Revolutionize value management: A mode towards sustainability*. International Journal of Project Management 25: 277.

Accenture Strategy, Circular Advantage Innovative Business Models and Technologies to Create Value in a World without Limits to Growth, https://www.accenture.com/t20150523T053139Z__w__/us-en/_acnmedia/Accenture/Conversion-Assets/DotCom/Documents/Global/PDF/Strategy_6/Accenture-Circular-Advantage-Innovative-Business-Models-Technologies-Value-Growth.pdf.

Adamczyk, J. (2001). *Koncepcja zrównoważonego rozwoju w zarządzaniu przedsiębiorstwem* [The concept of sustainable development in business management]. Wydawnictwo AE w Krakowie, Kraków, p. 122.

Adamczyk, J. and Nitkiewicz, T. (2007). *Programowanie zrównoważonego rozwoju przedsiębiorstw* [Programming sustainable development of companies], PWE, Warszawa, p. 27.

Ahvenniemi, H., Huovila, A., Pinto-Seppä, I. and Airaksinen, M. (2017). What are the differences between sustainable and smart cities? Cities. 60: 234–245. doi:https://doi.org/10.1016/j.cities.2016.09.009.

Antikainen, M., Uusitalo, T. and Kivikytö-Reponen, P. (2018). Digitalisation as an Enabler of Circular Economy, Procedia CIRP 73, pp. 45–49, 10.1016/j.procir.2018.04.027, https://doi.org/10.1016/j.procir.2018.04.027.

Arias, P.A., Villegas, J.C., Machado, J., Serna, A.M., Vidal, L.M., Vieira, C., Cadavid, C.A., Vieir, S.C., Ángel, J.E. and Mejía, Ó.A. (2016). *Reducing Social Vulnerability to Environmental Change: Building Trust through Social Collaboration on Environmental Monitoring*, American Meteorological Society, DOI: 10.1175/WCAS-D-15-0049.1.

Arnold, G. (2010). Inwestowanie w wartość, Wydawnictwa Profesjonalne PWN, Warszaw, p. 20.

Bachman, J.E. and Black, Wright P. (1997). *In Search of Shareholder Value*, Price Waterhouse, Pitman, p. 76.

Barnett, M. and Duvall, R. (2005). Power in Global Governance. Cambridge University Press, Cambridge.

Beulque, R. and Aggeri, F. (2016). *Circular Business Model Innovation: Key Patterns and Challenges to unleash recycling value creation potential.* EGOS, Naples, Italy. <halshs-01290810>.

Borys, T. (1998). *Jak budować program ekorozwoju, Informacje ogólne* [How to build an eco-development program, General information]. Regionalny Ośrodek Ekorozwoju Fundacji Karkonoskiej w Jeleniej Górze, Warszawa – Jelenia Góra, p. 11.

Brown, B. J., Hanson, M. E., Liverman, D. M. and Merideth, R. W. (1987). Global sustainability: Toward definition. Environmental Management, 11: 713–719.

Burchell, J. (2008). *The Corporate Social Responsibility Reader.* Routledge, Taylor & Francis Group, London and New York, pp. 111–118.

Burnham, P., Kent, R., Macartney, J. and Redfern, V. (2003). *Sustainable Land Management P568. Centre for Development, Environment and Policy*, University of London, London, Rydén, P. (2008) Sustainable Land Management—An Approach to Growth in Natural Resource Based Economic Sectors Why and How? World Bank, Washington DC.

Cerrato Garcia, E. (2024). *Navigating the landscape of EU sustainable finance regulation: challenges and imperatives to enforce sustainability financial promises in the EU through harmonized private enforcement mechanisms.*

Chodyński, A. (2012). *Środowisko naturalne jako interesariusz w koncepcjach odpowiedzialnego biznesu*, Przegląd Organizacji, Nr 1 (864): 21–25, https://doi.org/10.33141/po.2012.01.06.

Climate-Smart Agriculture, Managing Ecosystems for Sustainable Livelihoods, https://www.fao.org/climatechange/29790-0178d452d0ca9af024aad1092d4b78b1d.pdf.

Dauvergne, P. and Lister, J. (2013). *Eco-business: A big-brand takeover of sustainability.* MIT Press.

den Hollander, M. and Bakker, C. (2016). Mind the Gap Exploiter: Circular Business Models for Product Lifetime Extension. In Proceedings of the Electronics Goes Green, Berlin, Germany, 6–9 September 2016; pp. 1–8.

Directive 2014/95/EU Of The European Parliament And Of The Council of 22 October 2014 amending Directive 2013/34/EU as regards disclosure of non-financial and diversity information by certain large undertakings and groups.

Dudycz, T. (2001). *Finansowe narzędzia zarządzania wartością przedsiębiorstwa* [Financial tools for company value management], Studia i Monografie Nr 143, Wydawnictwo Akademii Ekonomicznej im. O. Langego we Wrocławiu, Wrocław, p. 9.

Epstein, M. J. and Buhovac, A. R. (2010). *Solving the sustainability implementation challenge.* Organisational Dynamics, 39: 307.

Geng, Y. and Doberstein, B. (2008). Developing the Circular Economy in China: Challenges and Opportunities for Achieving 'Leapfrog Development', International Journal of Sustainable Development & World Ecology", 15: 231–239.

Grudzewski, W.M., Hejduk, I.K., Sankowska, A. and Wańtuchowicz, M. (2010). *Sustainability w biznesie czyli przedsiębiorstwo przyszłości, zmiany paradygmatów i koncepcji zarządzania* [Sustainability in business, the company of the future, paradigm shifts and management concepts]. Wydawnictwo Poltext, Warszawa, p. 27.

Guinée, J.B. (2002). Handbook on life cycle assessment: operational guide to the ISO standards. In Book Review: The Second Dutch LCA-Guide; Springer Science+Business Media B.V.: Dordrecht, The Netherlands, pp. 311–313.

Hawken, P. (1993). *The Ecology of Commerce.* Weidenfeld&Nicolson, London, p. 12.

Hrasky, S. (2011). Carbon footprints and legitimation strategies: symbolism or action? Accounting, Auditing & Accountability Journal, 25: 174–198.

https://www.consilium.europa.eu/pl/policies/paris-agreement-climate/],[https://eur-lex.europa.eu/legal-content/EN/TXT/PDF/?uri=CELEX:22016A1019(01).

https://www.startus-insights.com/innovators-guide/climate-tech-trends-innovations/#:~:text=Top%2010%20Climate%20Tech%20Trends%20%26%20Innovations%20in,...%208%208.%20Carbon%20Data%20%26%20Analytics%20.

Ituarte-Lima, C. and Mares, R. (2024). Environmental democracy: Examining the interplay between Escazu Agreement's innovations and EU economic law. Earth System Governance, 21, 100208. https://doi.org/10.1016/j.esg.2024.100208.

Jabłoński, A. (2008). *Modele biznesu w sektorach pojawiających się i schyłkowych, Tworzenie przewagi konkurencyjnej przedsiębiorstwa opartej na jakości i kryteriach ekologicznych* [Business models in emerging and declining sectors, Creating a competitive advantage of the company based on quality and ecological criteria]. Wydawnictwo Wyższej Szkoły Biznesu w Dąbrowie Górniczej, Dąbrowa Górnicza, p. 19.

Jabłoński, A. (2010). *Modele zrównoważonego biznesu a koncepcja VBM oraz CSR, Przedsiębiorstwo Przyszłości* [Sustainable business models and the concept of VBM and CSR, Company of the future]. Kwartalnik Wyższej Szkoły Zarządzania i Prawa im. Heleny Chodkowskiej, Nr 2(3): 42

Jeżowski, P. ed. (2002). *Zarządzanie w sektorze publicznym – rozwój zrównoważony – metody wyceny* [Public sector management – sustainable development – valuation methods]. SGH, Warszawa, p. 47.

Johannesburg Summit (2002). World Summit on Sustainable Development, UN.

Johnston, R. (1966). *Technical Progress and Innovation*, „Oxford Economic Papers", 18(2): 158–176.

Kamela-Sowińska, A. (1996). Wartość firmy [Company value], PWE, Warszawa.

Keating, M. (1994). Globalny program działań [Global Action Program], Agencja Informacyjna "GEA", Warszawa, p. 13.

Kidd, C. (1992). The evolution of sustainability. Journal of Agricultural & Environmental Ethics, % (1): 1–26.

Kołodko, G.W. (2008). *Wędrujący świat* [The Wandering World]. Prószyński i S-ka, Warszawa, pp. 270–271.

Korhopnen, J., Honkasalo, A. and Seppälä, J. (2018). Circular Economy: The concept and its limitations. Ecolog. Econ. 143: 37–46.

Kotler, Ph. (1998). *Konkurencyjność a charakter społeczny, Organizacje przyszłości* [Competitiveness and social character, Organizations of the future]. eds. Hasselbein F., Goldsmith M., Beckhardt R., Wydawnictwo Business Press, Warszawa, p. 22.

Kotler, Ph. and Caslione, J. A. (2009). *Chaos, zarządzanie i marketing w erze turbulencji* [*Chaotics: The Business of Managing and Marketing in the Age of Turbulence*]. MT Biznes Sp. z o.o.

Kuciński, K. (2009). *Strategie przedsiębiorstw wobec wymogów zrównoważonego rozwoju* [Strategies of companies towards sustainability requirements]. SGH, Warszawa, p. 18.

Kumari, S. *Clean Energy: Key to a Sustainable Future*, ENVIS RP: Geodiversity & Impact on Environment Vol. 25(3).

Lash, J. and Wellington, F. (2007). Competitive advantage on a warming planet. Harvard Business Review, 85(3): 94–102.

Laubscher, M. and Marinelli, T. (2014). Integration of Circular Economy in Business. In Proceedings of the Conference Going Green—CARE INNOVATION, Vienna, Austria.

Li, F. (2017). *The digital transformation of business models in the creative industries: A holistic framework and emerging trends. Technovation*, https://doi.org/10.1016/j.technovation.2017.12.004.

Lieder, M. and Rashid, A. (2015). Towards circular economy implementation: a comprehensive review in context of manufacturing industry. J. Clean Prod 2015/115, pp. 36–51.

Lifset, R. and Graedel. T.E. (2001). *Industrial Ecology: Goals and Definitions*, w: R.U. Ayres, L. Ayres (eds.), Handbook for Industrial Ecology, Edward Elgar, Brookfield, pp. 3–15.

Linder, M. and Williander, M. (2015). Circular Business Model Innovation: Inherent Uncertainties. Bus. Strateg. Environ.

Manninen, K., Koskela, S., Antikainen, R., Bocken, N., Dahlbo, H. and Aminoff, A. (2018). Do circular economy business models capture intended environmental value propositions?. Journal of Cleaner Production 171, pp. 413-422, https://doi.org/10.1016/j.jclepro.2017.10.003.

Mokhlesian, S. and Holmén, M. (2012). *Business model changes and green construction processes. Constr. Manag. Econ*. 30: 761–775. [CrossRef].

Morin, R.A. and Jarrell, S.L. (2001). *Driving Shareholder Value: Value- Building Techniques for Creating Shareholder Wealth*, McGraw-Hill, New York.

Mousavizadeh, S. (2024). ESG factors in sustainable project management in wind power parks: problems and best practices, case studies from Finland and Iran.

Nußholz, J. L. K. (2017). *Circular Business Models: Defining a Concept and Framing an Emerging Research Field*, Sustainability, 9, 1810; doi:10.3390/su9101810, p. 12.

Piontek, B. (2002). *Koncepcja rozwoju zrównoważonego i trwałego Polski* [The concept of sustainable development of Poland], Wydawnictwo Naukowe PWN, Warszawa, p. 29.

Poskrobko, B. (1997). *Teoretyczne aspekty ekorozwoju* [Theoretical aspects of eco-development]. Wydawnictwo Ekonomia i Środowisko, Nr 1.

Poskrobko, B. (1998). *Podstawy polityki ekologicznej* [Environmental policy foundations], [w:] Górka, K., Poskrobko, B. and Radecki, W. (eds.). Ochrona środowiska, *Problemy społeczne, ekonomiczne i prawne* [Social, economic and legal issues], PWE, Warszawa, p. 75.

Preston, F. and Lehn, J. (2017). *A Wider Circle? The Circular Economy in Developing Countries*. Chatham House, London.

Roos, G. (2014). *Business model innovation to create and capture resource value in future circular material chains*. Resources 3: 248–274. [CrossRef].

Sauvé, S., Bernard, S. and Sloan, P. (2016). Environmental sciences, sustainable development and circular economy: Alternative concepts for trans-disciplinary research, Environmental Development, 17 (2016): 48-56, s.53.

Schulte, U.G. (2013). *New Business Models for a Radical Change in Resource Efficiency, Environmental Innovation and Societal Transitions*, 9.

Schumpeter, J.A. (1912). *The Theory of Economic Development, Transaction Publishers*, New Brunswick, New Jersey.

Skoczylas, W. (ed.). (2007). *Determinanty i modele wartości przedsiębiorstw* [Determinants and value models of companies], PWE, Warszawa, p. 15.

Smith, C.N. and Lenssen, G. (2009). Odpowiedzialność biznesu, teoria i praktyka [Corporate Responsibility, Theory and Practice], Wydawnictwo Studio EMKA, Warszawa, p. 274.

Spieth, P. and Schneider, S. (2016). *Business model innovativeness: designing a formative measure for business model innovation*. J. Bus. Econ. 86: 671–696.

Subramaniam, M., Iyer, B. and Venkatraman, V. (2018). *Competing in digital ecosystems*, Business Horizons. https://doi.org/10.1016/j.bushor.2018.08.013.

Svenson, G., Wood, G. and Callaghan, M. (2010). *A corporate model of sustainable business practices: An ethical perspective*. Journal of World Business 45: 338.

Szablewski, A. (ed.). (2000). *Strategie wzrostu wartości firmy* [Company Value Growth Strategies], Poltext, Warszawa, p. 15.

Technical Facilities Solutions "Building Management Systems" http://www.tfsmanagement.com/Building-Management-Systems.php.

Thompson, V.A. (1969). *Bureaucracy and Innovation*, University of Alabama Press, Tuscaloosa.

Tridech, S. and Cheng, K. (2008). *Low Carbon Manufacturing: Characterization, Theoretical Models And Implementation*, The 6th International Conference on Manufacturing Research (ICMR08) Brunel University, UK, 9-11th September 2008.

Unece Nexus, Sustainable Mobility and Smart Connectivity, Geneva, 2021, https://unece.org/sites/default/files/2021-04/2015779_E_web.pdf.

Velthuis, L.J. and Wesner, P. *Value Based Management, Bewertung, Performancemessung Und Managemententlohnung mit ERIC*, Schaffer Poeschel Verlag Stuttgart, p. 95.

Visser, W. (2010). *CSR.2.0. The Evolution and Revolution of Corporate Social Responsibility. In*: Pohl, M. and Tolhurst, N. 9eds.). *Responsible Business: How to Manage a CSR Strategy Successfully*, Wiley, London.

Visser, W. (2011). *The Age of Responsibility: CSR 2.0 and the New DNA of Business*, Wiley, London.

Wamane, G.V. (2023). *A "new deal" for a sustainable future: enhancing circular economy by employing ESG principles and biomimicry for efficiency*. Management of Environmental Quality: An International Journal.

Yaashikaa, P.R., Senthil Kumar, P., Saravanan, A., Karishma, S. and Rangasamy, G. (2023). A biotechnological roadmap for decarbonization systems combined into bioenergy production: prelude of environmental life-cycle assessment. Chemosphere 329: 138670. https://doi.org/10.1016/J.CHEMOSPHERE.2023.138670.

Zalega, K. (2003). *Systemy Corporate Governance a efektywność zarządzania spółką kapitałową* [Corporate governance systems and the effectiveness of capital company management], SGH, Warszawa, p. 27.

3

Systematic Literature Review

Description of the Climate Responsibility Research Landscape in the Context of Business Model Design from the Perspective of Management, Digital Transformation and the Technological Development of Companies

3.1 Introduction

The aim of this chapter is to present the results of a systematic literature review in relation to the characteristics of the climate responsibility research landscape in the context of business model design from the perspective of management, digital transformation and the technological development of companies. The scope of the chapter includes issues related to the sources of the concept of climate responsibility and the development of its theory along with the essence of climate responsibility in business models. The subject of the chapter concerns a multidimensional cognitive perspective aimed at shaping the principles of climate responsibility both in reference to theoretical cognition and implementation aspects.

3.2 Methodological assumptions of the systematic literature review

For the purpose of conducting comprehensive and multidimensional research and a scientific discussion on the dependence on shaping the principles of climate responsibility from the perspective of business models of companies embedded in the logic of digital and climate transformation at the levels of business, economy and society, a study was conducted by means of a systematic literature review.

DOI: 10.1201/9781003474401-3

To this end, relevant scientific articles and grey literature were used. The systematic literature review resulted in final conclusions and a report produced in accordance with the PRISMA (Preferred Reporting Items for Systematic Reviews and Meta-Analyses) standard (Tricco et al. 2018). The research gap is to identify the extent to which the theory of climate responsibility is combined with the concept of business models embedded in the logic of digital and climate transformation at the levels of business, economy and society. The relevant literature has not yet addressed this subject so broadly and multidimensionally, although it seems astonishing and surprising considering the logical, methodical and conceptual potential of the assumptions of the concept of corporate social responsibility and its extension to various components of management of various types of organizations. The research procedure comprised the following steps:

1. Gathering the relevant literature on social responsibility, climate responsibility and business models, as well as the principles of Digital Transformation and Climate Transformation.
2. Conducting a systematic literature review by means of the Web of Science and Scopus database.
3. Developing a report for systematic literature reviews and meta-analyses – application of the PRISMA standard.
4. Analyzing findings based on the results of research conducted as part of the method of the systematic literature review.

In the relevant literature, the concept of "climate responsibility" is currently not fully characterized and described. The aim of this systematic literature review was to examine whether and to what extent it occurs alongside the concepts of business model and climate transformation and digital transformation. Do the scientific achievements accumulated in the global scientific literature support the processes of studying and shaping the principles of climate responsibility from the perspective of business models of companies embedded in the logic of digital and climate transformation at the level of business, economy and society? Based on two leading databases of scientific publications, i.e., Web of Science and Scopus, quantitative results in terms of the common occurrence of these concepts in the scientific literature were obtained when searching for the terms "climate responsibility" and "business models" in the category of business, management and accounting.

The databases were searched for the following terms:

- climate AND responsibility and business AND model and climate AND transformation AND digital AND transformation

In the first stage of the study, the following results were obtained:

- Scopus database: 2 items
- WoS database: 10 items

The subsequent stage of the study consisted of removing duplicates, as a result of which 11 literature items remained (one article was repeated in both databases). In the next step, the analysis was conducted by means of the included/excluded method. This process uses the Rayyan tool, which supports researchers in conducting systematic literature reviews. Based on these studies, the PRISMA model was used and a report on the results of the systematic literature review was prepared in this form. The Figure 4 shows the PRISMA report for the given terms.

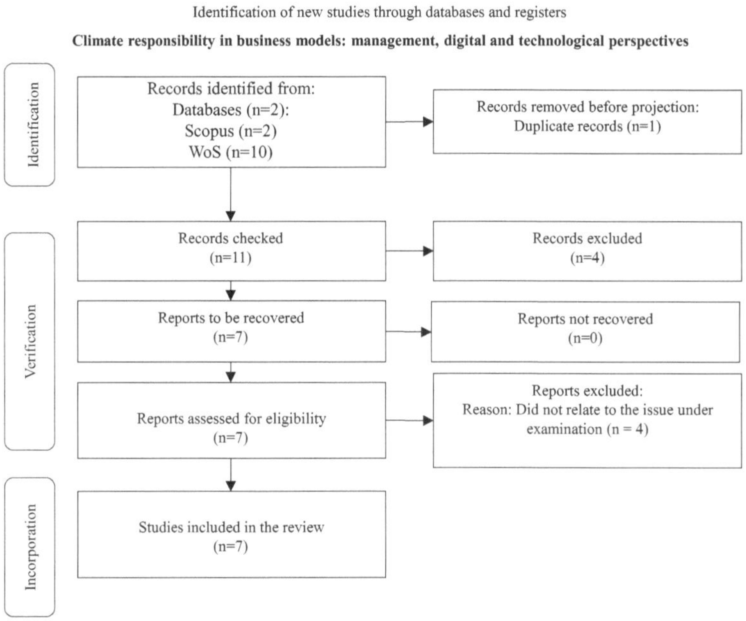

Figure 4. Report on the methodology of the systematic literature review. Source: Own study.

For articles that were included, a semantic map was prepared by means of VOSviewer.

In the Figure 5 semantic map related to the concept of climate responsibility includes concepts related to sustainability, business education, global competencies, global citizenship, global mindset and empirical study. Surprisingly, there is no connection with the concept of the business model, which is confirmed by the innovative layout of the monograph and its special interpretation in the context of the adopted mechanisms of strategic management. Linking the principles of climate responsibility with the concept of business models is therefore a new concept which is cognitively interesting and generates new prospects for the development of the economy and business in terms of dynamically occurring climate change and the adaptation thereto by various types of organizations.

3.3 Sources of the concept of climate responsibility against the background of strategic determinants of building business models

In order to be able to build the proper development potential of climate responsibility in many organizations, it is necessary to understand its deep and important strategic message, seeing it as an obvious and strong social and economic benefit. Only climate-sensitive and properly aware managers and their competent staff are able

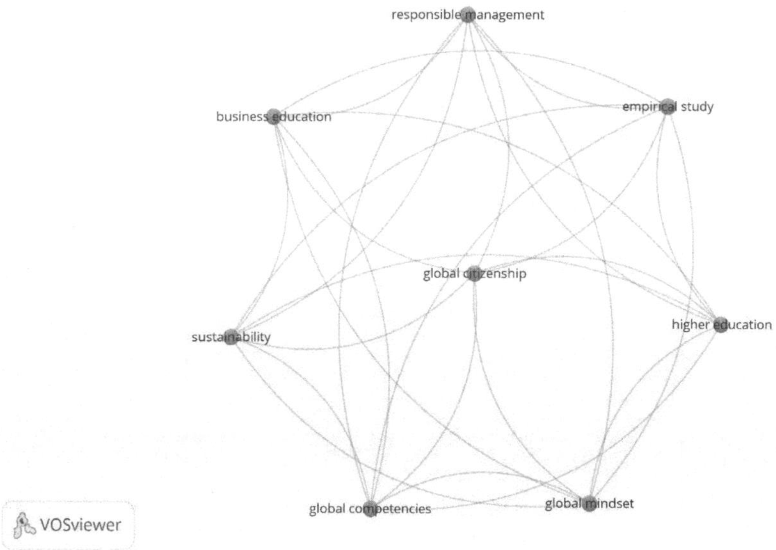

Figure 5. A semantic map of key concepts related to climate responsibility prepared by means of VOSviewer. Source: Own study.

to take advantage of the opportunities resulting from this substantively strong phenomenon at the global level. Despite the often antagonistic approach of numerous populists, not only at the political level, to the issue of climate change, many entrepreneurs and managers set themselves the goal of market leadership based on fulfilling climate criteria in production processes, products and services that fulfill climate requirements. Then, the source of the concept of climate responsibility is properly understood and interpreted self-awareness and sensitivity of various types of communities as well as business itself. It results directly from the understanding of dynamically progressing anomalies and climatic dysfunctions that already strongly interfere in every sphere of human life, along with a number of threats appearing with this area. There is a fundamental distinction between long-term climate change trends and short-term weather anomalies (extremes). In both cases, it is crucial to understand and scientifically describe the phenomena occurring here. In science, "anomaly" means a deviation from the average. In climatology, this is usually a deviation from the multi-annual average—from the selected reference period, usually 30 years. Indicating the anomaly values should also indicate which reference period it has been determined for. Some anomalies turn into climatic extremes, i.e., very strong, often uncontrolled phenomena. Local communities recognize these risks and point to the need to halt these adverse events and express content aimed at triggering all decision-making processes by different groups of decision-makers towards proactive action to halt dynamic and adverse climate change. The issue of meteorological and climatological extreme phenomena (events) and their actual and potential consequences in the natural and social environment is already frequently covered in the literature, which is largely related to the significant impact of these phenomena on the functioning of societies and the predicted changes in the frequency (mainly the increase) of these events. Effective and efficient management of the complex climate security ecosystem as a source of climate responsibility is also crucial here. Therefore, climate responsibility itself is not only related to understanding the place and role of climate responsibility in individual decision-making mechanisms, but is also a strategic factor that often determines the life and health of many people around the world. Because the main recipients of this responsibility are people as an entity threatened by the adverse consequences of climate change. Hence, the adopted logic of climate responsibility should be revealed in the business models used by companies. Their properly defined configuration should have a strong climatic basis built in the designated genotype of a climate-responsible organization. The canvas of such a business model should be built from components which strongly express climatic values.

3.4 Development of the theory of climate responsibility and its verification versus the construction of the business models of companies

Theory or practice, practice or theory? This repeatedly asked question is also relevant to the issue of climate responsibility. Irrespective of the order, it should be emphasized that only a scientific approach and scientific knowledge allows us not only to understand, but above all to describe, climate problems at local and global levels. A proper, scientific, substantive approach supported by the results of scientific research can give the proper communication message to even the greatest opportunists and opponents of the climate combat and stopping its abnormal changes. The concept of corporate social responsibility has been developing for many years, taking different contexts, forms and interpretations which result primarily from the changing paradigms of management against the background of real changes at the level of the socio-economic ecosystem in local and global terms. During this time, various types of changes were created, and crises and phases of dynamic development and growth occurred. Table 8 presents the evolution of corporate social responsibility.

It should be noted that in the course of the development of the concept of corporate social responsibility, the climate problem has not been solved to this day; on the contrary, it is highly complicated from political, social and economic perspectives. Scientific theories as products created in scientific processes and specific mechanisms of their construction and use perform various socio-cultural, economic and psychological functions. This also applies to the concept of climate responsibility. The development of the theory of climate responsibility can be based

Table 8. The evolution of corporate social responsibility.

No.	The evolution of corporate social responsibility
1.	Corporate social management in the 1950s and 1960s
2.	Social reaction of companies in the 1960s and 1970s
3.	Corporate social integrity in the 1980s and 1990s
4.	Business as a global citizen in the 1990s and 2000s

Source: Own study.

on climate change models and the use of their potential to effectively and efficiently manage various types of organizations. Climate models describe physical processes in the atmosphere, oceans and cryosphere (snow and ice cover). They are used to predict the climate in the future and to study it on the basis of historical data. The theory of climate responsibility is not yet well established and sufficiently described in the relevant literature. In particular, it should be noted that it is interdisciplinary and is linked to many scientific fields and disciplines. For example, it covers the disciplines of management and quality sciences, climate sciences, natural sciences and climate engineering or climate security sciences. Therefore, extensive debates take place in various fora on who is responsible for climate change, whether it is the state, companies or individual users of natural space. The Table 9 describes the responsibility for climate change at the level of the state, the company and members of society.

Table 9. Description of responsibility for climate change.

No.	Responsible for climate change	Description of Responsibility
1.	Country	The state plays a very important, even fundamental, role in shaping climate policy and strategy at the level of the whole country. On the one hand, it implements global and European regulations on climate responsibility into its legislative system and, on the other hand, it creates its own, internal formal and legal regulations in this area. The state develops the mechanisms of climate education by creating and building strategic information, knowledge and data on climate in various areas of human existence of a given state. It upholds the law related to the fulfilment of climate and pro-ecological criteria.
2.	Company	It complies with climate law and ecological criteria. It builds its climate policy and strategy within the organization. It develops an organizational climate conducive to building a climate culture within the organization. It builds the climate competence of staff and develops climate capital at all levels and functions of the organization. It is open to the needs and expectations of customers who see the principles of climate responsibility and promotes pro-climate behavior among all groups of its stakeholders and/or actors of its business network. It recognizes the pro-climate behavior of its employees and motivates them to develop further in-depth principles of climate responsibility.
3.	Member of Society	They adhere to the climate law of the country where they live and exist. They recognize climate problems and, as a social unit, respect the principles and criteria of climate responsibility on their own and in their living processes. They educate the closest members of their local community in the area of direct pro-ecological and climate behavior.

Source: Own study

Since climate change is a global phenomenon, size adjustment implies that individuals would blame larger entities, such as multinational corporations, rather than individuals. Size adjustment may also mean that society would consider that targeting larger entities is likely to be more effective than targeting individuals (Swim and Geiger 2021). Therefore, it is crucial to appoint appropriate groups of climate stakeholders. It is also essential to indicate an innovative approach to the subject of climate responsibility, focusing on the aspects of improving the management of organizations in terms of the climate, climate security management, climate engineering, climate science, etc., towards the dynamic development of climate responsibility principles in the context of applicable systems, processes and models as great social vulnerability to climate change occurs. Then, community-oriented responses include activities whose main or partial purpose is to improve community resilience (Sovacool and Linnér 2016). Figure 6 presents embedding the new theory of climate responsibility against the background of management and quality sciences - squares instead of circles.

The author's original approach to shaping and understanding the new theory of climate responsibility reveals the far-reaching interdisciplinarity of this issue, in which it should be assumed that the theory of management and quality science is embedded primarily in the field of science theory, and more broadly in the theory of management science. An important part of this theory is the theory of climate security science, which fits into the theory of security. In addition, issues in the area of management and quality science theory include the theory of climate science, which also includes the theory of climate engineering and other theories of science, e.g., psychology (the human factor), ergonomics (the working environment), biology and others.

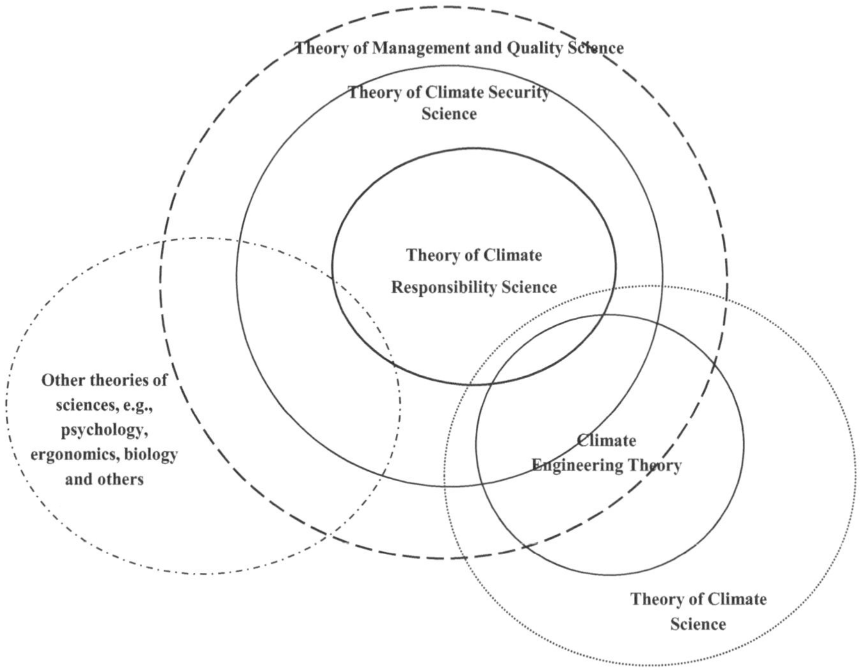

Figure 6. Embedding the new theory of climate responsibility against the background of management and quality sciences—squares instead of circles. Source: Own study.

3.5 The essence of climate responsibility in the business models of companies

The essence of climate responsibility stems directly from a lack of responsibility for the climate in a narrow and broad sense. A narrow understanding of climate irresponsibility applies to every individual who understands or does not understand climate risks. A broad understanding results from the institutional approach at the level of countries, regions, cities and companies and various types of other organizations. In order to be able to achieve success in this core area for the world, consistency should occur between a narrow and a broad understanding of this strategic problem. Therefore, climate responsibility can have individual, group, and institutional dimensions. All dimensions are crucial for the climate security of countries and regions and should be considered as such. Climate security is currently a strategic priority because the number of catastrophic events is growing at a geometric rate. This translates directly into human losses and huge financial damage. Moreover, the ongoing dynamics of events and threats make it difficult to maintain an appropriate level of stability of the economy and business, which turns into accelerating negative situations that often require immediate, systemic crisis management. The social and economic costs of these phenomena are very high. Therefore, the essence of climate responsibility is in the needs and expectations of numerous social groups who wish to have even a minimum certainty of a stable life and professional existence. This is about the present and the future of future generations, which is not fully understood and accepted among various individualized interest groups. Therefore, the essence of climate responsibility has both philosophical and operational dimensions associated with direct climate-related decision-making. In philosophical terms, this is aimed at a full understanding of the depth of the climate issue and, in operational terms, it is determined by a set of executive actions that neutralize climate risks at the level of the economy, business and society. It should be remembered at this point that very often it is the individual or man himself who determines only his individual behaviors, which are transferred to other individuals. This is the case for managers and entrepreneurs, where their decision-making processes determine the behavior of others who are influenced by them.

3.6 Conclusions

A systematic literature review in the form of the theoretical and practical approach to the issue of climate responsibility is fundamental to understanding its place and role in the business, economic and social existence of companies and other types of organizations. Understanding the concept of climate responsibility at the level of global literature allows for a multi-criteria assessment of the impact of climate responsibility on the proactive or reactive actions of various types of actors for climate change and adaptation to them. At the same time, it allows for different opinions related to the benefits and costs of developing not only climate responsibility in the economy but also at the level of individual countries, regions and cities.

References

Sovacool, B.K. and Linnér, B.O. (2016). *The Political Economy of Climate Change Adaptation*. Palgrave McMillan, Houndmills.

Swim, J.K. and Geiger, N. (2021). *Policy attributes, perceived impacts, and climate change policy preferences*, Journal of Environmental Psychology 77: 101673, https://doi.org/10.1016/j.jenvp.2021.101673.

Tricco, A.C., Lillie, E., Zarin, W., O'Brien, K.K., Colquhoun, H., Levac, D., Moher, D., Peters, M.DJ., Horsley, T., Weeks, L., Hempel, S., Akl, E.A., Chang, C., McGowan, J., Stewart, L., Hartling, L., Aldcroft, A., Wilson, M.G., Garritty, C., Lewin, S., Godfrey, C.M., Macdonald, M.T., Langlois, E.V., Soares-Weiser, K., Moriarty, J., Clifford, T., Tunçalp, Ö. and Straus, S.E. (2018). *PRISMA Extension for Scoping Reviews (PRISMA-ScR): Checklist and Explanation*. Ann Intern Med. 2; 169(7): 467–473. doi: 10.7326/M18-0850. Epub 2018 Sep 4. PMID: 30178033.

4

Corporate Climate Responsibility and the Business Models of Companies

4.1 Introduction

The aim of this chapter is to present the place and role of corporate climate responsibility against the background of the development, design and implementation of the business models of companies. The scope of the chapter includes the issue of corporate climate responsibility in the mechanisms of shaping business models and climate strategies, the consistency of business, economic and social perspectives in fulfilling the principles of climate responsibility and the principles of climate risk management of companies considered from the perspective of climate responsibility. The subject of the chapter concerns a deep look at the issues of corporate climate responsibility determined by the dimension of configuration of business models and the climate strategies of companies open to the consistent development of business and the economy for positive climate change.

4.2 Corporate climate responsibility in the mechanisms of shaping business models and climate strategies

The observation of climate change by various types of entities shapes the mechanisms of perception of this issue and determines the ways to take action in this respect or conversely the lack of such a response, which is also considered an action in management processes. However, it is worth being aware that only a decision based on facts is credible and true. Climate research is much more complex than mathematical modelling of the economy, and can thus be seen as more authentic. Responsibility for the climate should be borne by every person and he/she is the main beneficiary of this. However, not everyone understands the threats and negative phenomena associated with the climate from the perspective of spatial perception of the reality around us. This applies, among other things, to the concepts of the weather

DOI: 10.1201/9781003474401-4

and the climate. The difference between the weather and the climate determines the proper understanding and interpretation of climate issues. Weather refers to the instantaneous state of the atmosphere at a specific place and time. Climate is a long-term weather pattern characteristic of a given region. This is what we can expect from the analysis of long-term weather data. The weather is dynamic and difficult to predict. It is constantly changing and is influenced by various factors. The climate is more stable and predictable. It is based on the analysis of long-term data and shows typical weather values for a given area. Climate change has a huge impact on our daily weather conditions. They are increasingly the result of human activity. Increased greenhouse gas emissions, deforestation and pollution have an impact on climate change and contribute to extreme weather events such as droughts, floods and hurricanes. Understanding the difference between the weather and the climate is important because it affects our daily lives and enables us to take action on climate protection. Knowledge of climate change and its impact on weather conditions allows us to make the right decisions that will help protect the environment and improve our quality of life. Weather is a dynamic and volatile phenomenon, involving many atmospheric factors. It includes parameters such as temperature, air humidity, wind speed, precipitation and more. The weather changes from hour to hour and can be forecast based on the analysis of meteorological data. Weather variability is part of the natural environment and affects our daily activities. Climate is a long-term weather pattern characteristic of a given region. It includes average values of temperature, precipitation, humidity and other atmospheric parameters. Climate stability is the maintenance of these typical values over a longer period of time and allows for the predictability of weather conditions. The analysis of long-term weather data is a crucial element in the study and understanding of the climate. Unprecedented climate change causes climate extremes on a large scale including, among other things, heat, droughts, fires, downpours and heavy rainfall, snowfall and snowstorms and storm surges, hurricanes and tornadoes. All this generates a whole package of negative events, often of a catastrophic nature. At that time, climatic phenomena are atmospheric phenomena, as well as related phenomena that pose a threat to the city's population, the natural environment, buildings and infrastructure and the economy. From this perspective, vulnerability to climate change and the adaptation potential of the region are also important. Climate vulnerability is the extent to which a city is affected by climatic events. It depends on the nature of the spatial structure of the city and its individual elements, taking into account the population living in the city, its characteristics and spatial distribution. Vulnerability is considered in the context of the impact of climatic events, which can be direct or indirect. Adaptation potential is the city's tangible and intangible resources that can be used to adapt and prepare for climate change and its impacts. Adaptation potential is created by financial resources, human resources, institutional resources, infrastructure resources and knowledge resources. Vulnerability to climate change is understood as the extent to which a city is unable to cope with the adverse effects of climate change. Vulnerability depends on the city's vulnerability to the adverse effects of climate change and its adaptive potential. From the perspective of the European Union, the EU Emissions Trading System is particularly important. The Directive establishes an Emissions Trading System (ETS) in the European Union (EU). It underpins the Union's policy for combating climate change and aims to reduce greenhouse gas emissions in a cost-effective and economically efficient manner. It is based on the cap-and-trade principle (Directive 2003/87/EC).

4.3 The consistency of the perspectives of business, the economy and society in adhering to the principles of climate responsibility

The consistency of perception and understanding and the logical justification of climatic phenomena resulting from professional scientific research gives rise to multidimensional discourses on the climate. A scientific discourse supported by evidence based on reality should be a fundamental component of management processes in the political, social and economic dimensions. Business, the economy and society are governed by their own laws. There may be different interpretations of the place and role of climate responsibility in business, the economy and society. This is related to the concept of adaptation to climate change. Adaptation (i.e., actions that reduce the damage caused by climate change) is widely recognized as one of the two pillars of climate action, alongside mitigation (i.e., actions that reduce the concentration of greenhouse gases that cause climate change). So far, actions in both pillars have been widely considered insufficient (Orlove 2022). Adaptation to climate change takes place at the level of states, companies and social units embedded in local regions. It has different dimensions and different resonances and climatic effects. In terms of evolution, adaptation should be understood as a form of natural selection, which is at the same time a biological process during which changes in the construction, structure and function of organisms occur. Evolution is caused by changes in the living environment of a species (Darwin 1866). Such discussions can be extrapolated and developed based on the issue of climate change. This allows for the development of a scientific approach to climate change and interpreting and inferring it in this way. This enables the perspectives of business, the economy and society on fulfilling the principles of climate responsibility to be explained based on facts and reliable assessments and opinions. Table 10 presents description of perspectives of business, economy and society. When trying to develop this analytical–conceptual problem, it should be noted that in the case of business, this perspective may be expressed in an understanding of the mechanisms and needs related to climate responsibility as well as a lack of understanding or decision-making neutrality in this area. This is determined by the issue of self-awareness and the sensitivity of managers as well as the time perspective of doing business at the level of a specific company. If decision-makers within the organization are guided by the perspective of long-term market existence and wish to care for future generations of humanity, specific decision-making processes which shape a positive impact on climate change at the level of this particular company can be expected, but if there is no such perspective, the results of these actions will not be satisfactory. Then, only a factor stimulating positive climate change at the level of the company will have formal and legal regulations that positively affect the climate, generated by country-specific requirements. The consistency of the business perspective at the internal level of the organization takes place by means

Table 10. The perspectives of business, the economy and society in adhering to the principles of climate responsibility.

No.	Perspective	Degree of compliance with the principles of climate responsibility
1.	Business	In business, the concept of ecological economics applies, which can be expanded to include the principles of the climate economy. One of the basic categories of ecological economics is natural capital, understood as the resources and values of the natural environment together with ecosystem services, including: • procurement services (such as food, water, fiber and fuel), • regulatory services (such as carbon sinks, mitigation of floods and waste treatment), • cultural services (such as spiritual values, aesthetic impressions, recreation), • supporting services (such as soil formation and nutrient recycling) (Gunesch 2018).
2.	Economy	The development of ecology economics, otherwise known as the green economy, is positively correlated with higher economic growth. A similar situation occurs in the relationship between the economy and the climate economy. The level of climate investments is increasing dynamically.
3.	Society	Public awareness of the risks associated with extreme climatic events and how to reduce their impact on the environment, including people, is increasing dynamically. Without systemic civic, environmental and energy transition education, the achievement of climate neutrality will be much more difficult.

Source: Own study.

of the so-called cohesion of the strategic hybrid composed of four core ontological entities of the organization, that is, processes, projects, strategy and the business model (Jabłoński 2015).

A company may gain a competitive advantage when an organization generates above-average results compared to its competitors. Often the source of this advantage is the organization's resources or access to them. In order for resources to be a source of competitive advantage, certain conditions must be fulfilled, namely that they must be valuable (create value), rare, difficult to imitate and well-organized (VRIO – value, rare, inimitable, organization) (Barney 1986, pp. 656–665). B.W. Husted and D.B. Allen defined corporate social strategy (CSS). In their model, the CSS strategy is parallel to the company's strategy. Both of these strategies contribute to building a competitive advantage (Husted and Allen 2011). The strategy of social responsibility should be equipped with attributes of climate management. Then, new strategic opportunities emerge in the area of building the potential to compete and create new sources of competitive advantage. A competitive advantage based on climate criteria can be a new value driver for many types of organizations in different sectors of the economy. The competition potential can be embedded in various components of climate management, which should be reflected in the business model used. Competitive position achieved through climate criteria and values can be a new driver for companies to achieve high efficiency and reliability. Compilation of climate competitiveness components presents Table 11.

It should be noted that the classic competitive advantage is being replaced by technological advantage in the era of dynamic development of the economy based on the development of digital technologies. Technological advantage is a situation in which a company has its own unique and innovative technologies and/or access

Table 11. Compilation of climate competitiveness components.

No.	Climate competitiveness component	Description
1.	Competition potential based on climate criteria and values	Potential revenue growth drivers • How will changes in customer demand patterns affect prices? • What percentage of climate-related costs will we be able to shift to customers? • How can we generate revenue streams from new low-carbon products? • What new forms of income (for example, carbon credits) will become available? • What are the risks of low-carbon substitutes? • What impact will weather patterns have on revenue? (Lash and Wellington 2007).
2.	Competitive advantage based on climate criteria and values	Climate competitiveness refers to a company's ability to operate in a low-carbon, sustainable economy.
3.	Competitive position based on climate criteria and values	A measure of competitive advantage achieved through climate criteria and building a climate value proposition.

Source: Own study.

to those new technologies that give it a significant and noticeable advantage over competitors with higher revenues and profits. Technological development is one of the most important factors currently affecting market competitiveness. Companies that do not keep up with changing market trends and are unable to implement new technologies quickly become unattractive to current and potential customers. What matters here is technological entrepreneurship. Typical features of companies based on new technologies include multi-localization with a strong globalization trend, specialization and playing the role of the so-called technology broker, and companies based on networks managing external and internal relationships (Onetti et al. 2023, pp. 337–368). As a result, companies close the existing technology gap, which is treated as the difference between the levels of technological potential of individual companies.

4.4 Socially responsible movements and strategic corporate climate responsibility

Pro-climate movements and actions are a very important link in the space of understanding climate change around the world. Proactivity in relation to climate policies and strategies should also be bottom-up and supported by the so-called third sector of the economy, i.e., various types of non-profit and not-for-profit organizations, including various types of associations and foundations. Nowadays, it is important to be aware that a climate stakeholder is no longer a silent stakeholder, as was the case in defining stakeholders in terms of environmental criteria. Scientists need to become involved in climate change; stakeholder involvement in building climate responsibility is not an esoteric, pleasant thing, but it is a fundamental requirement to create information that is actually useful (Jebeile and Roussos 2023). Social movements in the climate field can create a positive message about the required actions of various actors of social and economic life for the climate. Unfortunately, they can also, to some extent, distort the real picture of climate problems by exaggerating effects that, for example, do not eventuate. The requirement is therefore to generate expectations of climate change based on hard evidence and facts. Climate movements are useful as long as they are credible and logically and scientifically justified. In addition, climate movements should be conducted while taking into account properly built relationships with other climate policy makers, countries and regions. They should also be open to a broad social dialogue based on the consensual development of solutions that are also relevant to economic and financial opportunities and to maintaining the continuity of business and the economy. It must be remembered that climate movements are particularly visible at the global level, as they usually seek to influence international climate negotiations of a strategic nature for the whole earth.

Climate activists are using key locations in cities to protest and express their demands and are starting to design their strategic activities in relation to local politics, without giving up on attempts to have a wide impact on the national and international policies of individual countries. This strongly visible trend may be very important for the further development of activism and its potential for social change in the area of climate responsibility.

4.5 Business climate neutrality as an insufficient attribute in a dynamic natural and socio-economic ecosystem

Climate neutrality can be understood as the relative equilibrium (zero balance) between emitted greenhouse gases and their storage or absorption by water reservoirs, forests or soils. Achieving climate neutrality is one of the biggest goals and challenges of climate protection. It is not sufficient for companies to generate climate neutrality because the degree of environmental impact and degradation may also be very high in a given region in historical terms. This is an evolutionary activity, which from the perspective of the ever-increasing dynamics of climatic anomalies may not give the result that is expected in the long term. Actions towards climate neutrality should be based on a carefully selected and monitored measurement system, the result of which is specifically defined target values of quantitative indicators with a specific time by which they should be achieved. This system should be tightly verified at specific time intervals based on the so-called milestones. Climate neutrality should therefore be the first step towards further challenges in the climate change space.

The policy conditions of the European Union shape the assumptions of climate neutrality of companies and regions. This includes, but is not limited to:

- EU legislation on the promotion of renewable energy sources,
- A share of renewable energy in the EU energy mix of up to 40% by 2030 – Fit For 55 concept,

- The basis for implementing the updated EU target of reducing greenhouse gas emissions by 55% by 2030,
- A number of legislative acts responding to the demands, including:
- Preserving the redistributive elements in the ETS in the form of a solidarity pool and increasing the Modernization Fund,
- Maintenance of free permits in the ETS,
- Retaining the criteria for allocating Member States' liabilities for sectors not covered by the ETS;
- The creation of a new fund to mitigate the risk of increasing energy poverty,
- The introduction of a border tax on emissions.

The so-called Green Deal is created as a result.

The benefits of the European Green Deal are illustrated in the Figure 7:

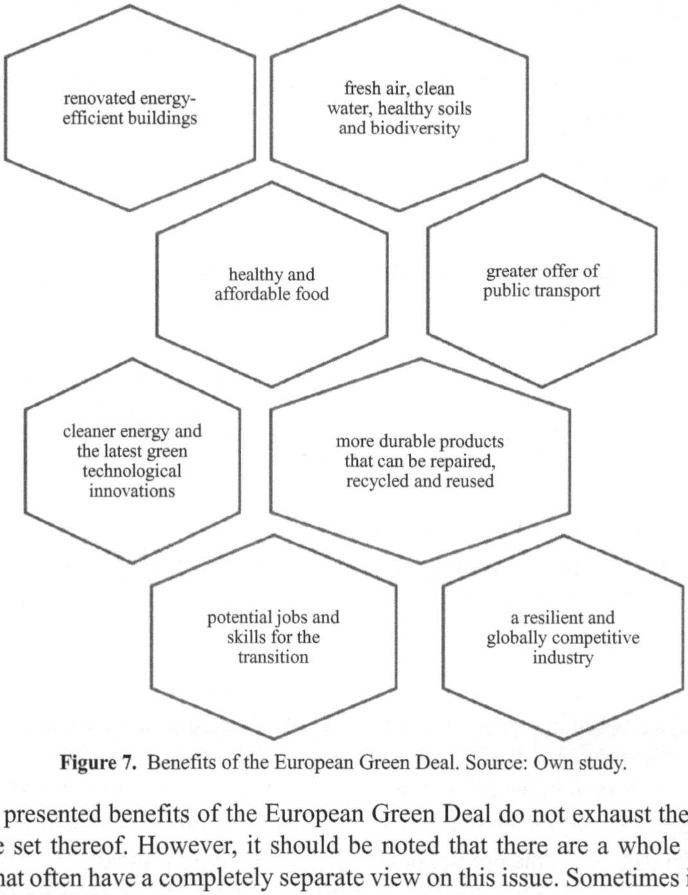

Figure 7. Benefits of the European Green Deal. Source: Own study.

The presented benefits of the European Green Deal do not exhaust the full and complete set thereof. However, it should be noted that there are a whole range of groups that often have a completely separate view on this issue. Sometimes it is even reduced to the so-called predatory climateism, characterized by a strong ideology, which is definitely detrimental to this concept that combats dynamically progressing adverse climate change.

4.6 Managing the climate risk of companies examined from the perspective of climate responsibility

Risks and threats are an integral part of management processes. Effective and efficient risk management is not easy. The strategic problem is related to risk perception, because each of us understands and perceives it differently. What may be a threat to one person may be a normal existential situation to another. Sometimes even a threat to one person can be a factor stimulating the activity of another person. Some can open their umbrellas in the rain and others can dance in the rain. Hence, the proper understanding of risk susceptibility as well as proper mitigation of threats is difficult. This is also the case for climate risks and threats. Climate risk management should be one of the core strategic factors for the development of companies. Catastrophic risk carries the aspect of a lack of knowledge of risk with it in the sense that governments and institutions have no experience in dealing with such incidents (Castellano 2010, pp. 391–415). Natural and anthropogenic processes are elements that change the energy balance of the Earth, that is, they are the driving forces of climate change. New climate technologies should open up market opportunities for wisely managed companies (Khan et al. 2024, pp. 770–772). In this case, it is important to reduce social vulnerability to environmental changes—building trust through social cooperation on environmental monitoring (Arias et al. 2016). Risk is examined from the perspective of the probability of the occurrence of a negative event generating a potential loss or damage, loss in quantitative terms and loss in financial terms. In the case of climate disasters, losses and damage are almost always significant. Hence, companies should satisfy the criteria dedicated to high-performance organizations and high-reliability organizations. Typology of selected climate risk categories shows Table 12.

Table 12. Typology of selected climate risk categories.

No.	Typology of selected climate risk categories
1.	Internal risk
2.	External risk
3.	Explicit risk
4.	Implicit risk
5.	Inherent risk
6.	Active risk
7.	Passive risk
8.	Functional risk
9.	Strategic risk
10.	Tactical and operational risk
11.	Personal risk
12.	Procedural risk
13.	Systemic risk
14.	Total risk

Source: Own study

Significant climate risk groups that need to be managed effectively and efficiently include systemic risks arising from extreme weather events (e.g., heavy rainfall, floods, droughts, heat waves, frost waves, strong winds), mortality and morbidity risks during heat waves, risk of environmental deterioration and income in rural areas, risk of food insecurity, risk of loss of marine, coastal, terrestrial and inland ecosystems, and biodiversity. Climate change poses the greatest threat to entities that do not intend to change their business strategy in the face of climate risk and choose the "business as usual" path (Henderson et al. 2018, pp. 1–39). This, by definition, triggers the process of generating various types of threats and risks. Risk refers to the probability of adverse events occurring in relation to the loss or damage they generate, losses in quantitative terms and damages in financial terms. Total probability creates certainty. Climate risk is a social risk. It often has a subjective or objective dimension. It all depends on the cognitive perspective of those who interpret it. This causes the so-called climate problem, which often gives rise to extreme assessments but also emotions. Therefore, such issues should be looked at in scientific rather than intuitive terms, for example. Therefore, the key in the assessment of climate risk is the adoption of appropriate boundary conditions for its identification. Climate risk must be managed in a systemic manner, where the management system is defined as an aggregated set of components in a designated function of the target, in this case aimed at increasing climate security by reducing the associated risks. Climate risk has a managerial, technological and economic dimension. Core global climate risks include, but are not limited to, the shrinking cryosphere, the rising sea level, increased warm extremes (e.g., the number of hot days and tropical nights, the frequency of heat waves), decreased cold extremes (e.g., the number of cold days and nights) and the increased concentrations of carbon dioxide in the atmosphere. Climate risks translate directly into macroeconomic indicators such as inflation, economic growth, financial stability and monetary policy transmission. As a result of climate change, new sources of financial risk have emerged, reinforcing the climate–financial correlation (Battiston et al. 2021). Climate risk is inextricably linked to financial risk (Giglio et al. 2021, pp. 15–36). Climate change generates additional threats to ecosystems, while the continuous degradation of nature significantly accelerates climate change. Climate risks can include climate anomalies and natural disasters. In technological terms, climate risk includes a gap in how new sustainable business models operate, e.g., in the energy sector, while examining their risk profile, especially in relation to the risk of regulatory changes (Leisen et al. 2019, pp. 865–878). It is crucial that companies of different types and profiles take action to integrate climate risk and its derivatives into their risk identification process. An understanding of the current and future risks of climate fragility is essential for effective climate fragility programming. This includes assessing the dimensions of resilience to these risks in order to identify potential entry points for interventions. In order to fully understand these factors of fragility and plan for future risks, considerations should be given to climate change issues in conflict assessments (Climate Fragility Programming Resource Guide 2024).

4.7 Global migration crises and climate responsibility at the level of countries, regions, cities and businesses

A crisis is an undesirable phenomenon of a continuous and often long-term and large-scale nature. This is the case with migration crises. Migration is the movement of a person or group of persons across an international border or within a country. It is a movement of the population, involving any type of movement, regardless of the distance travelled, the participants and the reasons for such action; it includes the migration of refugees, displaced persons, labor migrants and persons moving for other purposes, including family reunification (Glossary on Migration 2011, pp. 62–63). In recent years, migration as a mass phenomenon has become not only a challenge for the national security of sovereign states, but also a subject of reflection on the humanitarian dimension of this very important issue. Migration crises on a global scale in relation to broadly understood climate responsibility at the level of states, regions, cities and businesses are a huge strategic problem. In addition to the internal dimension, the migration crisis also has implications for the foreign and security policy of other countries in their external relations. Migration crises have been caused, among others, by the escape of many people from climate-vulnerable areas. Climate responsibility therefore has a very difficult dimension at the level of human existence on earth in many regions of the world. Therefore, the climate migration policy and its core strategic assumptions become crucial. At this point, it is worth noting first of all what criteria are followed by climate immigrants fleeing from areas threatened by weather anomalies, extremes and climate dysfunctions and climate disasters:

- saving human life and health,
- access to water and food and thus lack of hunger,
- the possibility of functioning in a stable natural ecosystem,
- access to work and acceptable existential conditions.

These criteria also determine positive or negative decisions taken by decision-makers in the areas where climate immigrants go as well as management processes in this very difficult topic from the perspective of properly understood climate responsibility at the level of states, regions, cities and companies.

4.8 Conclusions

The principles of corporate climate responsibility should be followed at different levels of the operation and exploration of business, the economy and society. It should be a platform for changing the polarization of business towards creating positive climate change. The search for solutions relevant to climate needs using the concept of business models and climate strategies should open a wide path to building climate knowledge platforms. This knowledge should be a guideline for the strategic behavior and thinking of business managers and policy makers of regions and cities. Only a systemic approach to shaping the principles of climate responsibility can give rise to changes aiming to reverse adverse climate trends. This requires a multidimensional discourse at the level of science and business practice.

References

Arias, P.A., Villegas, J.C., Machado, J., Serna, A.M., Vidal, L.M., Vieira, C., Cadavid, C.A., Vieir, S.C., Ángel, J.E. and Mejía, Ó.A. (2016). *Reducing Social Vulnerability to Environmental Change: Building Trust through Social Collaboration on Environmental Monitoring*. American Meteorological Society, DOI: 10.1175/WCAS-D-15-0049.1.

Arias, P.A., Villegas, J.C., Machado, J., Serna, A.M., Vidal, L.M., Vieira, C., Cadavid, C.A., Vieir, S.C., Ángel, J.E. and Mejía, Ó.A. (2016). *Reducing Social Vulnerability to Environmental Change: Building Trust through Social Collaboration on Environmental Monitoring*, American Meteorological Society, DOI: 10.1175/WCAS-D-15-0049.1.

Barney, J.B. (1986). *Organizational Culture: Can It Be a Source of Sustained Competitive Advantage?*" Academy of Management Review, 11: 656–665.

Battiston, S., Dafermos, Y. and Monasterolo, I. (2021). *Climate risks and financial stability*. "Journal of Financial Stability", 54. https://doi.org/10.1016/j.jfs.2021.100867.

Castellano, G. (2010). *Governing ignorance: emerging catastrophic risks - industry responses and policy frictions*. The Geneva Papers. 35: 391–415.

Climate Fragility Programming Resource Guide, U.S. Department of State Bureau of Conflict and Stabilization Operations, U.S. Agency for International Development Bureau for Conflict Prevention and Stabilization, https://www.state.gov/wp-content/uploads/2024/10/Planning-Resource-Guide.pdf.

Darwin, K. (1866). *On the Origin of Species by Means of Natural Selection, or the Preservation of Favoured Races in the Struggle for Life*. Londyn, s. 61, http://darwin-online.org.uk/Variorum/1866/1866-71-c-1861.html.

Directive 2003/87/EC establishing a scheme for greenhouse gas emission allowance trading within the European Union.

Giglio, S., Kelly, B. and Stroebel, J. (2021). *Climate finance. Annual Review of Financial Economics*, 13 (3): 15–36.

Gunesch, K. (2018). *A critical comparison of environmental economics and ecological economics, combined with the creative value system of degrowth, for a macroeconomically efficient and equitable sustainable development*, Review of Socio-Economic Perspectives, 3. Issue: 2, DOI: 10.19275/RSEP049.

Henderson, R.M., Reinert, S.A., Dekhtyar, P. and Migdal, A. (2018). *Climate Change in 2018: Implications for Business*. Harvard Business School, 1–39. www.hbsp.harvard.edu.

Husted, B.W. and Allen, D. B. (2011). *Corporate Social Strategy*, Cambridge University Press, Cambridge.

International Organization for Migration, Glossary on Migration, 2nd Edition. 2011, nr 25, Geneva, pp. 62–63.

Jabłoński, A. (2015). *Spójność hybrydy strategicznej w środowisku sieciowym* [Consistency of the strategic hybrid in the network environment], Difin, Warszawa.

Jebeile, J. and Roussos, J. (2023). Usability of climate information: Toward a new scientific framework, WIREs Clim Change.14: e833. wires.wiley.com/climate change, https://doi.org/10.1002/wcc.833.

Khan, F., Byers, E., Carlin, D. et al. (2024). *Science-based principles for corporate climate transition risk quantification. Nat. Clim. Chang*. 14: 770–772. https://doi.org/10.1038/s41558-024-02067-2.

Lash, J. and Wellington, F. (2007). *Competitive Advantage on a Warming Planet*, Harvard Business, http://courseresources.mit.usf.edu/sgs/geb6930/module_3/read/competative_advantage.pdf.

Leisen, R., Steffen, B. and Weber, C. (2019). *Regulatory risk and the resilience of new sustainable business models in the energy sector.* Journal of Cleaner Production, pp. 865–878. https://doi.org/10.1016/j.jclepro.2019.01.330.

Onetti, A., Zucchella, A., Jones, M. and McDougall-Covin, P. (2023). *Internationalization, innovation and entrepreneurship: business models for new technology-based firms*, Journal Management Government 16: 337–368.

Orlove, B. (2022). *Annual Review of Environment and Resources*. pp. 47: 535–81, https://doi.org/10.1146/annurev-environ-112320-095719.

5

Climate Services as a Tool for Building Corporate Climate Responsibility

5.1 Introduction

The aim of the chapter is to present climate services as new paths towards an organizational and technological revolution in the area of climate management. Climate services embedded in business models can strongly counteract adverse climate change. The scope of the chapter includes setting the principles of the assumptions of the climate transformation of the economy and society. The subject of the chapter refers to the aggregated set of selected climate services. They are applied in various sectors of the economy and society, often changing the logic of climate impact.

5.2 Climate services in shaping the business models of companies

The products and services offered by businesses and other organizations can have a direct or indirect impact on the climate and climate change. If all products and services are produced and used in a natural ecosystem, which at the same time is a socio-economic ecosystem, each product and service affects the climate and its changes in this sense. Therefore, the development of a national strategic framework for the development of climate services should be a priority for any climate-conscious state. The development of climate services can be based on an iterative and dynamic collaboration process that engages in dialogue between different types of expertise to consistently integrate climate services into decision-making for a climate-resilient society, treating users as experts in their own fields of expertise. The climate services market can be expected to evolve in the near future as more investment is made and the effects of climate change become a daily reality to be faced. In order to understand where demand will come from in the future, it is important to understand which sectors need help at the moment. For example, it is important to build

DOI: 10.1201/9781003474401-5

knowledge systems and actions on adaptation to climate change in the EU. It should be noted that, ideally, climate services should be jointly developed by scientists and stakeholders who work together to identify user decisions and needs. However, while climate services have been well developed on a regional and local scale, relatively little attention has been paid to them on a global scale. It is worth asking the question in the context of climate change of whether this is a "forecast" or a "projection".

Often, the data set reflects the diversity of climate services, but it also allows for the identification of certain attributes that were more common than others. For example, the most common type of service reported was seasonal climate information provided by national meteorological services, in cooperation with research institutes, to agricultural operators via the Internet. Therefore, a very important component of climate services, to which particular attention should be paid, is their maturity and the degree of their impact in stopping negative climatic phenomena. This applies to both public and private, commercial climate services. Integration between the public and private sectors is very important in this area. Currently, the dynamically developed mechanisms of climate knowledge management are an integrating factor. Moreover, as regards public climate services, they are financed by taxpayers for the benefit of those taxpayers, i.e., the society of a given region or country. The situation is different in the case of private climate services, which are commercial services and should therefore generate at least a reasonable profit. Then, an economically justifiable climatic value is created.

Climate change should never be overlooked in decision-making processes at different levels of these decisions. The understanding of climate change should have a spatial dimension in the logic of knowing reality and should never be trivialized. Any such action simplifies this problem and reduces it to insignificance, which can cause far-reaching and uncontrolled effects in the future that can no longer be reversed. Dynamic phenomena currently taking place in the space of the economy, society and business shape the mechanisms of decision-making by various types of decision-makers. This applies to the levels of the European Union (Lipcaneanu and Amanatidis 2023; EU 2020), countries and/or different types of organizations. The European Union is one of the leading major economies when it comes to counteracting greenhouse gas emissions. In 2020, EU greenhouse gas emissions fell by 31% from 1990 levels and reached their lowest level in 30 years, exceeding the EU's Kyoto Protocol target of a 20% reduction by 2020. In 2019, the Commission presented the concept of the European Green Deal and is now proposing measures to increase the ambition of reducing greenhouse gas emissions by 55% by 2030 and to decarbonize the EU economy by 2050, in line with the commitments made under the Paris Agreement. Today, decision-making processes are significantly hampered, especially in the context of layered overlaps between individual crises with significant impact. They form a kind of ecosystem that is a sophisticated environment of numerous crises, including the climate crisis. This environment is composed of, among others, economic, social, military, health, migration and energy crises, and the climate crisis plays an important role. This crisis, which is manifested, among other things, through a wide stream of uncontrolled threats such as earthquakes, floods and volcanic eruptions, is unfortunately dynamic in many cases as a result of many years of human intervention in the natural environment. The result of this ongoing process is significant climate warming, various types of hurricanes, heavy rains with strong storms, forest fires, tsunami and others. This makes one completely helpless in the face of such threats or the accumulation thereof in many cases. From such a cognitive perspective, there is a need to change the logic and methodology of understanding these threats as well as trying to effectively manage these threats. The issue of climate services as a kind of a new tool used not only to neutralize the adverse impact of various factors on the environment, but above all to reverse the negative polarization of the impact of the economy and society, is at the forefront. This can be used to create a positive impact on the natural ecosystem, for example, by means of business models and organizational strategies that shape positive climate change. This is especially important for dynamic and proactive managers focused on creative thinking in the context of climate change.

The following research questions can be asked:

1. What value and benefits can climate services generate for the economy and society in terms of halting, reversing or eliminating adverse climate change?
2. What solutions and approaches can be used to create climate services in various types of economic sectors through the concept of business models?
3. What is the place and role of business models in shaping climate services by using the implemented principles of climate transformation?

Appropriate substantive and formal supervision of the development of climate services by various types of public and private entities is crucial here. The research questions posed have, on the one hand, postulative significance and are, on the other hand, explanatory and focused on achieving specific research results. These types of questions are important especially from the perspective of generating a positive economic and climate effect, the recipients of which are all groups of society, business and the economy. Based on research into climate services, nine definitions of climate services were selected, illustrating their importance to a wide range of economic sectors. They set out a core descriptive attribute included in the definitions in the Table 13.

When conducting a broad and multidimensional analysis of the presented definitions of climate services, it can be concluded that they can have two supporting fundamental cognitive functions:

Cognitive Primary Function: Climate services (usually as new and innovative climate services) are created by various groups of their creators through their built-in functions. These applications can directly or indirectly affect the state or climate change (the original assumption is that it should be positive). Thus, new climate services are dynamically and positively changing the climate (a proactive approach).

Cognitive Secondary Function: As a result of this climate (climate change), climate services are created and proposed as a response to stopping the most frequent adverse climate changes. Thus, the climate and its current state provokes an appropriate decision-making response in the form of the creation and implementation of climate services (a reactive approach).

When analyzing the presented definitions of climate services, their practical character related to management and its processes and functions is at the

Table 13. Review of key definitions of climate services.

No.	Definition of Climate Service	Core Attribute Definition
1.	Transformation of climate-related data, together with other relevant information, into tailor-made products such as projections, information, trends, economic analyses, assessments (including technological ones), advice on best practices, the development and assessment of solutions and any other climate-related services that may be useful to the general public (European Commission 2015).	Data transformation, best practices, useful for society
2.	Climate services are defined by the "WMO Global Framework for Climate Services" as "providing climate information in such a way as to aid decision-making" (Hewitt et al. 2012).	Information, decision-making
3.	Definition of Climate Europe: "Climate service is the provision of climate information to assist decision-making. The service must respond to user needs, must be based on scientifically credible information and expertise, and requires appropriate engagement between the users and providers" (climateurope 2020).	Information, decision-making, user needs
4.	Climate services can be seen as a form of sustainable innovation that "takes into account environmental, social and economic considerations in their development and use" (Larson 2007).	Sustainable innovation, development, use
5.	Climate services are a specific type of knowledge-based activity, knowledge-intensive business services (KIBS), i.e., characterized by non-financial services requiring a high level of knowledge due to the high density of human capital (Brenner et al. 2017).	Knowledge, non-financial services
6.	These services are designed to facilitate climate adaptation, disaster risk mitigation and reduction, which are widely recognized as important sustainability challenges in both rich and poor countries (Asrar et al. 2012).	Climate adaptation, disaster risk, sustainability
7.	Climate services include those features that incorporate them into business models (Larosa and Perrels 2017).	Functions, business models
8.	Climate services can be defined as decision support tools developed based on the process of transforming climate information into appropriate advisory services that assist in decision-making by individuals and social organizations. The climate service provides tailored, relevant and useful advice to policymakers and vulnerable communities, based on available climate information (for example, forecasts). The service requires an appropriate involvement in the creation of advice relevant to the user's needs, to assist him/her in making decisions in conditions of uncertainty and to facilitate and set the direction of early action and readiness. The service also means the effectiveness of the access and delivery mechanism. First of all, a climate service responding to the needs of users is necessary (Tall and Njinga 2013).	Decision-making, climate information transformation, advisory services
9.	Climate services embody the production, explanation, and transfer of climate research into useful information for climate-related decision-making (Vaughan et al. 2018).	Production, research transfer, decision-making

Source: Own study.

forefront here. Their main goal is to support decision-making processes for various groups of decision-makers, including managers at various levels. In this interpretive context, it is important to define the broad context of the application of climate services in business, the economy and society. The Global Climate Services

Framework (GFCS) was launched by delegates from 155 countries in September 2009 "to strengthen the production, availability, supply and application to the climate of science and foresight-based services". Since then, climate services around the world (World Meteorological Organization) have been provided in multifaceted terms. This is evidenced, inter alia, by forecast-based humanitarian funding mechanisms and the growing interest in promoting the use of climate services. This is supported, among other things, by:

- climate risk management in agriculture,
- natural disaster risk reduction,
- implementation of health protection mechanisms,
- changes in water management, energy management and many other sectors vulnerable to climate change.

Climate services exist in the complex public–private space and cover a number of climate-vulnerable sectors (Brasseur and Gallardo 2016). They contain a broad set of actions, including climate data, information and knowledge (e.g., supporting adaptation, mitigation and natural disaster risk management). They also cover the demand for these service and the assessment thereof.

Global decisions at the highest level using climate information include:

- multilateral climate policy decisions – many countries decide for themselves on climate policy measures affecting many or all of them;
- multi-regional/localization decisions – individual decision-makers (e.g., countries, companies, etc.) exposed to climate risk in numerous countries and diverse locations around the world (Bisaro et al. 2021).

In this interpretation, a multidimensional discussion among scientists around the world about the differences between climate services (CS) and a climate services information system (CSIS) takes place. Information and knowledge may concern, inter alia, data on climate variables (adaptability to climate change), the format of these data (e.g., tools, manner of visualizations), the intended uses of this information (e.g., raising awareness or support for specific decision-making processes), the way information is delivered (online media, television, radio and other media, including social media) and the interactions related to communication and dissemination (e.g., helpdesks).

Climate services are crucial to risk management. They make it possible to build political responsibility and take advantage of the various opportunities offered by a changing climate. However, there is evidence of unequal quality among many of the climate services available on the market. Doubts arise as to the quality of the data processing processes (such as bias correction techniques, the quantification of uncertainties and their suitability for a specific purpose) and, consequently, the quality of the service provided and the implementation of a given climate service. This raises several core issues:

- poorly conducted climate services can lead to erroneous or even harmful climate actions;

- the lack of trust in the suitability of climate information for a specific purpose, which is fostered by the poor quality of services, is a huge obstacle to climate action. This delays the implementation of major agreements and policies, such as the Paris Agreement and the European and American Green Deal, among others (Baldissera Pacchetti and St. Clair 2023).

Climate services are also increasingly seen as an important means of effectively supporting individuals, communities and organizations to adapt to the long-term impacts of climate change (Boon et al. 2022). Long-term climate information (mainly related to decades and centuries of time scales) is typically associated with long-term investments and planning decisions, such as large infrastructure and national development plans (Jones et al. 2015a, 2015b). It should also be remembered that investing time and resources in understanding the local political context and engagement with national and local partners can strongly help promote more effective communication and wider use of climate information (Jones et al. 2017). Climate services potentially create a socially relevant and profitable market for consultancy services and meteorological research in institutes. To develop this new market, a change is required in which attention paid to the collection of observational data and the creation of accurate models is complemented by demand for these data and a focus on its users (Visscher et al. 2020). Climate service providers are located in the middle of the value chain, sitting in the so-called "Valley of Death" (Osawa and Miyazaki 2006). Research into climate services argues, inter alia, that climate services could strongly benefit from the adoption of a service-dominant logic (co-creation of value in the service-dominant logic) (Grönroos 1999). The importance of subjective user experiences is recognized and the position of users as value co-producers is strengthened (Alexander and Dessai 2019). There is one specific aspect of climate services that has not yet been fully achieved. Great importance is attached to ensuring the quality and standards of climate services, in which we believe that significant progress will take place in the coming years. Standards and processes for data quality assurance and meteorological systems already exist, but not for specific climate services. Quality assurance and these standards are needed in the supplier–user interface. They ensure that these services correspond to the agreed levels and convey sufficient clarity, reliability and authority. They also include rules on the use of services to avoid inappropriate or unintended abuse, as well as to build bilateral trust between suppliers and recipients (Hewitt and Stone 2021).

5.3 Understanding the importance, place and role of climate services in shaping climate responsibility

The appropriate understanding of the importance, place and role of climate services in shaping climate responsibility is determined by numerous interpretative factors and attributes. Currently, a very heated discussion is taking place in the scientific and practical field of how to understand climate services, how to aggregate them substantively in the areas of business and society, how to possibly expand them in terms of product services and how to embed them against the background of the public sector and private sectors in the economy. In addition, it is difficult to determine which types of climate information and decision support are considered important. In this approach, a discourse on the level of theory and practice of defining the concept of climate service and its features, attributes and parameters of logical interpretation occurs. This transforms into a new climate value offered by these new service providers. Moreover, it is necessary to pay attention to the climatic factors taken into account by managers and entrepreneurs in their decision-making from strategic to tactical and operational levels. From this perspective, it is also important to refer to climate information compared to other input data and policy priorities. The space and time scale of climate knowledge and information (e.g., short-term forecasts, seasonal forecasts and long-term forecasts) as the most relevant for business, but taking into account sectoral conditions in terms of the climate, is also important here. Another aspect in such theoretical and practical discourse related to climate services should be the elimination of any disinformation at the level of decision uncertainty or contradictory information in the socio-economic ecosystem. Then, a multidimensional and multi-criteria assessment of the value and benefits in the shape and image of climate services can be conducted. Achieved and potentially generatable benefits should go far beyond the horizon of current thinking about climate change, creating broadly understood innovative, rare and unique climate solutions such as green urban spaces, better living conditions and better conditions in the professional and private spheres. Managers can then make more informed and beneficial climate decisions in various fields of business and human life with the help of these newly created and modern climate services. In order to be able to do so effectively and efficiently, new challenges or obstacles encountered in the use of creative climate services need to be fully tackled. The determination of their cost-generating nature against the background of economic benefits should also become the goal of scientific and practical debate.

5.4 Overview of the types of selected climate services and their application in different sectors of the economy

Climate services are increasingly expanding in terms of theory and application and thus generating a stream of new, innovative, smart climate values. This powerful technological power should be increasingly recognized and used appropriately in both the public and private sectors. Public and private consortia with complex financial systems built with equity and external capital should be built at the level of climate products and services. Therefore, different types of climate products and services should be properly developed, tested and implemented. In the context of the application of the climate services offered, taking into account their various definitions, attention is paid to their interdisciplinary and widely developed substantive nature. This is due to the wide impact of climate change on various spheres of human existence around the world. From the perspective of such an impact, a number of services are included both in the analog and digital system generating climatic values. They have a strong social and economic impact and are already recognized by many stakeholder groups in both the public and private sectors. They underpin new business and market opportunities, including as a strategic response of various types of entities to the numerous deepening crises.

The Table 14 presents an overview of the types of selected climate services.

Table 14. Overview of the types of selected climate services.

No.	Type of climate service	Short description
1.	Maps and applications that provide real-time information on the state of the climate	These types of services are focused on providing a whole range of climate information to people in a given area. For example, tourists very often go hiking in the mountains and download maps and weather data, e.g., an upcoming storm or rain in the area, where they are in real time with forecasted changes in the near future.
2.	Climate change databases	These types of services are related to the collection of aggregated data and their processing. They include historical and current data together with the possibilities of their processing in the area of various analytical needs.
3.	Offering dedicated climate and agronomic models	These types of services are a significant source of strategic information for farmers, who can plan the processes of managing crops of various products in different climatic areas and managing, for example, large animal farms. Thanks to this, they can purchase raw materials and conduct planning processes for the development of crops and breeding.
4.	Services generating climate data and forecasts	These types of services are used to conduct a number of processes in the economy and society. For example, they constitute knowledge necessary for the location of various types of investments, e.g., construction of new production plants or public infrastructure facilities. They can be strategic information for tourists planning trips to various places in the world.
5.	Expert services including data analysis and model generation to develop contextual weather forecasts and advice	Such services are often necessary for the comprehensive management of given regions or specific countries in terms of development, investment processes or those resulting from the current existential needs of a given region or country.
6.	Integrated counseling covering decision-making on the anticipation and mitigation of climate change	Such services are often a system supporting strategic decisions for both business entities and public management decision-makers.
7.	Interdisciplinary services in the field of management, climate engineering together with climate data management	These types of services are used, for example, in the processes of water and sewage management or forest management. It is then possible to plan and manage these specific issues, including those of a natural character, effectively and efficiently.
8.	Tools for creating climate projection graphs	These planning and forecasting services provide strategic knowledge for development, investment or change management processes in a given thematic area related to the functioning of regions or business or society and related to climate change dynamics.
9.	Climate risk modelling services focused on the identification and measurement of climate-related threats and opportunities	Such services are used, inter alia, to manage investments in areas threatened by strong climate impact. They are also used to build optimal systems for crisis management and the management of major accidents or climate disasters, together with their socio-economic and technological analysis.

Table 14 contd. ...

...Table 14 contd.

No.	Type of climate service	Short description
10.	Systems supporting decision-making processes in the area of climate change	Such services support different types of decision-making processes in the area of climate and its impacts, including impacts at different levels of decision-making in society, business and the economy.
11.	Integrated regional research and development programs in the field of climate change	These types of services based on scientific solutions resulting from in-depth scientific research provide strategic knowledge for the development and growth of various regions of the world.
12.	Common climate change adaptation systems and programs creating climate knowledge	Such services provide a common platform for the exchange of knowledge and technology. They are an interdisciplinary process of shaping and interpreting various research results, observations or behaviors of different communities in the area of climate management.
13.	Supporting local adaptation to climate change through participatory Geoweb systems	These types of services provide current information for operational decision-making processes for both business and society.
14.	Workshops and training to raise public and business awareness about climate change and to promote tools and solutions used in processes supporting climate change	These types of services provide a platform for shaping attitudes that support and strengthen climate awareness and sensitivity. By building the transfer of knowledge, they create the principles of climate responsibility in business, the economy, society and politics.
15.	Seasonal climate and agricultural forecasts dedicated to specific locations	Thanks to this type of service, farmers can plan harvests and determine purchase and investment needs. They can also analyze the short- and long-term agricultural potential in agricultural and livestock production areas.
16.	Organizing dedicated events, festivals, study meetings promoting climate services in traditional and online form, raising awareness and climate sensitivity and aimed at creating interdisciplinary networks of contacts and relationships in the field of climate change	Such services create pro-climate attitudes and provide platforms for shaping coherent views and interpretations related to the climate. They also bring people with similar views and values together. They provide a source of knowledge of current trends, needs and social, economic and technological problems in the areas of climate management.

Source: Own study.

When conducting a wide analysis of selected climate services, it is worth paying attention to the fact that they include climate change forecasting services using specially selected mathematical and statistical modelling. They also include various types of advisory services, expert or digital platforms supporting climate information in various fields of their operation. It is noted that, de facto, they now affect almost all sectors of the economy and their various dimensions. This is reflected in the Table 15, which presents selected climate services that have a positive impact on climate change.

The aggregated data in the table above clearly show that climate services have huge pro-development intellectual climate potential. They can not only reduce the adverse impact on climate change (stop it) but also reverse this negative polarization

Table 15. List of selected climate services positively influencing market changes in various sectors of the economy.

No.	Climate service and its brief description	Economic sector
1.	The assumption of climate risk by the insurance sector along with the possibility of paying dedicated compensation for extreme weather events. This also includes the management of so-called transition risks, i.e., risks associated with actions taken by state authorities and regulators to, inter alia, reduce greenhouse gas emissions.	Climate and catastrophe insurance
2.	Conducting tests and simulations of economic climatic conditions and analysis of scenarios for dedicated portfolios of projects, products, services and business models.	Financial services
3.	Delivering transport and logistics services taking into account the principles of the European Green Deal, which aims to deliver a multidimensional stream of actions aimed at transforming Europe into a modern, resource-efficient and competitive green economy. It is characterized by: (1) net-zero GHG emissions in 2050, (2) the separation of economic growth at the level of countries, taking into account the use of resources, (3) equal and cohesive living standards in all regions of Europe.	Transport and logistics services
4.	Climate reporting of companies listed on the WSE with recommendations of the Task Force on Climate-Related Financial Disclosures (TCFD) with the implementation of the principles of responsible investment (PRI).	Mainly banking and insurance services
5.	Expert carbon footprint calculation covering emissions of carbon dioxide, methane, nitrous oxide and other greenhouse gases (GHG). The carbon footprint is expressed in CO2 equivalent (CO2e), which is a universal unit of measurement indicating the global warming potential (GWP) of each greenhouse gas, expressed in terms of GWP of one unit of carbon dioxide. This includes the total sum of greenhouse gas emissions caused directly or indirectly by an entity, which may be a company, a product, a service, an event, a city, a municipality or a state, but also a private individual.	Expert services
6.	Breeding new varieties of crops resistant to water scarcity and thermal stress, growing winter crops, not leaving unsown fields, breeding new breeds of animals with increased resistance to heat stress and more frequent pressure changes, and adapting breeding to the ability to produce fodder in changed climatic conditions, protecting soils from erosion and improving fertility, monitoring fertilization processes, introducing new methods to control pests and diseases of plants and animals, protecting water in rural areas (small retention, green belts and field trees, agrotechnical treatments, collecting water from roofs and hardened irrigation sites) and increasing farmers' awareness of climate change and its effects (Ministry of Agriculture and Rural Development).	Agricultural services
7.	Climate technologies, regardless of sector-specific circumstances, may concern: (1) technologies directly reducing or eliminating emissions to air, soil, water and resource use, (2) technological solutions helping adapt to the effects of climate change in various areas of industry and services, (3) technological solutions to raise awareness of the impact of climate change on the basis of a range of data collected and processed.	Climate technology design services

Table 15 contd. ...

...Table 15 contd.

No.	Climate service and its brief description	Economic sector
8.	Building green infrastructure as an ecological and spatial concept to promote the health and resilience of ecosystems, support biodiversity conservation and create benefits for people by promoting ecosystem services such as climate change mitigation, the creation of key habitats for flora and fauna and ecological connectivity.	Construction services
	Production of hydrogen as an alternative energy source. The objectives of the hydrogen economy include: (1) Implementation of hydrogen technologies in the energy sector – reducing the intensity of the use of fossil fuels and reducing the state's dependence on fuel imports, which will improve energy security in the long term. (2) The use of hydrogen as an alternative fuel in transport, especially in urban, road and maritime transport, and in the longer term also in aviation, which will reduce CO_2 emissions by replacing conventional fuels. (3) Supporting the decarbonization of industry – the use of low-carbon hydrogen produced using RES will allow for a significant reduction in greenhouse gas emissions. (4) Hydrogen production in new installations – taking into account the direction set by the European Green Deal and the EU Hydrogen Strategy, Poland's objective for hydrogen production will be to ensure the right conditions for the deployment of installations for the production of hydrogen from low- and zero-emission sources. (5) Efficient and safe distribution of hydrogen – the strategy aims to achieve hydrogen transport capacity by 2030 (gas mixture containing approx. 10% of gases other than natural gas) through gas networks. (6) Creating a stable regulatory environment – a regulatory framework allowing for the use of hydrogen as an alternative fuel in transport, the foundations for the functioning of the hydrogen market and the development of a legislative hydrogen package – rules defining the details of the functioning of the market, implementing EU law in this regard and implementing a system of incentives for the production of low-carbon hydrogen.	Hydrogen production and support services

Source: Own study.

to create positive climate change in individual sectors of the economy. As a result, they can change their size and character. The optimal instrument that can be applied to this process is the concept of business models.

5.5 Conclusions

Climate services should be accessible to all key users and should often be made available free of charge or cheaply. They may include, inter alia, various expert analyses, mainly on multidisciplinary climate services tailored to specific decision-making situations. Integrated multi-data sharing practices should be clearly defined together with the sharing of climate services and climate policy services between different actors. They should also include interdisciplinary management and climate engineering services. This requires the use of "prosumer" knowledge through the creation of effective platforms for the exchange of climate knowledge and information. The knowledge of individual prosumers then increases the overall intellectual potential of functioning climate communities.

References

Baulenas, E., Bojovic, D., Urquiza, D., Terrado, M., Pickard, S., González, N. and St. Clair, A.L. (2023). *User Selection and Engagement for Climate Services Coproduction*, American Meteorological Society. For information regarding reuse of this content and general copyright information, consult the AMS Copyright Policy (www.ametsoc.org/PUBSReuseLicenses)., DOI: 10.1175/WCAS-D-22-0112.1.

Bisaro, A., Hinkel, J., Le Cozannet, G., van der Pol, T. and Haas, A. (2021). *Global Climate Services: A Typology of Global Decisions Influenced by Climate Risk*. Front. Mar. Sci. 8: 728687.doi:10.3389/fmars.2021.728687.

Bray, D. and von Storch, H. (2009). *"Prediction" or "Projection"?: The Nomenclature of Climate Science*. Science Communication, 30(4): 534–543. https://doi.org/10.1177/1075547009333698.

Golding, N., Lambkin, K., Fischer, A.M., Wilson, L., Hygen, H.O., De Troch, R., Hama, A.M., Dyrrdal, A.V., Jamsin, E., Termonia, P. and Hewitt Ch. (2025). *Developing national frameworks for climate services: Experiences, challenges and learnings from across Europe*, Climate Services, Volume 37, 100530, https://doi.org/10.1016/j.cliser.2024.100530.

Panenko, A., George, E. and Lutoff, C. (2021). *Towards the development of climate adaptation knowledge-action systems in the European Union: An institutional approach to climate service analysis*, Climate Services 24, 100265, https://doi.org/10.1016/j.cliser.2021.100265.

Tarta, S., Groth, M. and Seipold, P. (2020). *Market demand for climate services: An assessment of users' needs*, Climate Services, Volume 17, 100109, https://doi.org/10.1016/j.cliser.2019.100109.

Vaughan C., Dessai S., Hewitt Ch. (2018). Surveying Climate Services: What Can We Learn from a Bird's-Eye View?, American Meteorological Society, DOI: 10.1175/WCAS-D-17-0030.1.

6

Climate Determinants of Socio-Economic Ecosystems in Strategic Terms

6.1 Introduction

The aim of the chapter is to present the mechanisms for constructing and implementing business models and strategies that positively change the climate in the context of functioning sectoral conditions in which given organizations exist. The scope of the chapter covers the determinants of strongly and moderately climate-influencing sectors in proactive and reactive dimensions of business models and strategies. The chapter also covers the characteristics of selected sectors in the context of the construction and implementation of business models and strategies that change the climate.

6.2 Socio-economic and natural ecosystems in climate change management processes

The proper and logically justified understanding, interpretation and perception of current dynamic natural and socio-economic ecosystems should be strongly supported by strategic reflection on the earlier years (the history of the climate and climate change). This is particularly important in the effective and efficient fight against climate change or related omissions, e.g., in relation to systemic action or the lack of systemic development of electricity systems built from an appropriate electricity mix, particularly including renewable energy sources.

These ecosystems are an obvious and natural market environment for doing business and developing the economy. This development is qualitative and creates an appropriate amount of quantitative growth. Development must then be fully compatible with growth. The climate change dynamics around the world coincide with the increasing dynamics of various types of crises, both at the level of the economy and society.

DOI: 10.1201/9781003474401-6

Both global decision-makers and local communities are increasingly aware of the climate risks directly or indirectly affecting different aspects of human life (Hewitt et al. 2021). Climate policy and adaptation to climate change is now becoming a strategic priority, and not only in Europe (Dolsak and Prakash 2018, pp. 43: 317–41; Orlove 2022, pp. 47: 535–81). Therefore, the decision-makers at different levels of management try to search for the best possible solutions in the management sphere and optimal solutions in economic terms and implement them into various areas of the socio-economic ecosystem in order to improve the quality of human life and the efficiency of business and the economy. This ecosystem is supported by important digitalization solutions (Kohtamäki et al. 2019, pp. 380–392). A crucial ontological entity currently at the disposal of managers in order to achieve the projected results of the organization is a strategy understood primarily as an aggregated and consistent set of long-term goals of the organization and the business model. The business model is usually a specific and unique configuration of tangible and intangible resources that generates a value proposition acceptable to customers. It is aimed at fulfilling the principles of optimal income generation logic implemented by means of the abovementioned sophisticated strategy.

In the context of dynamic climate change, a new taxonomy of business models is emerging. Business models shaping climate components include sustainable business models (Jabłoński 2013a), circular business models (Mentink 2014), social business models (Yunus et al. 2010, pp. 308–325), progressive business models (Rochlin 2021), inclusive and adaptive business models (Rosenstock et al. 2020, pp. 76–81), green business models (Jabłoński A. 2013b) and technological business models (Baden-Fuller and Haefliger 2013, pp. 419–426; Jabłoński A. 2015). As a result, a new type of business model emerges, i.e., climate-changing business models, which are also strongly linked to climate services (Larosa and Mysiak 2020). A climate-changing business model is a business model that aims to create climate benefits and values for different groups of the ecosystem of society, the economy, and business. The outcome of this model is usually climate services that have a positive influence on the climate and generate a simultaneous climatic and economic effect. The platform for creating such business models is organizational culture shaped in organizations based on the assumptions of climate corporate responsibility (Jabłoński 2023). Climate strategies are also referred to in the broad scientific conceptual framework (Paul et al. 2017, pp. 50–70). They are supported by scientific solutions related to digital transformation (Fischer et al. 2020).

It should be noted here that both the strategy and the business model are very strongly dependent on the defined sectoral conditions in which the organization operates and competes on the market with other entities. Sectoral conditions determine the appropriate shape and dimension of the set, long-term goals of the organization and condition the canvas of the business model dedicated to a given sector. It must have an appropriate configuration structure with its corresponding, mutual interfaces creating the expected integrity and functionality of the business model. Key attributes of a given sector include its competitiveness, stability and the appropriate life cycle state (emerging, mature, declining). The attractiveness of the sector and its economic and technological potential are also important. The strength of the sector's impact on its internal aspects and the strength of its impact on other sectors of the economy also determine its potential. It can be attractive or repulsive, particularly in terms of potentially investable financial, social, innovative and intellectual capital. In such a cognitive space, every manager must be fully aware that the determinant of well-structured strategies and business models is a proper understanding and interpretation of sectoral conditions that strongly determine the strategic trajectory for the development and growth of wisely managed companies. Operating in an environment of numerous crises, managers try to orient their strategies and business models and navigate them accordingly in the direction of using these crises not only to eliminate threats, but above all, to search for market opportunities that open new space for generating positive, free cash flows. Undoubtedly, climate change opens up such a path. The climatic conditions of socio-economic ecosystems of different regions of the world determine the effective and efficient mechanisms of management of various types of organizations, especially against the background of sectoral conditions. From such a cognitive perspective, determining the place and role of the climate in business and the economy are the core components of strategic management. They are embedded in the decision-making processes of organizations aimed at a proper understanding of climate dysfunction and its impact on the economic, social and ecological performance of modern companies. In this approach, the key is to understand the climate as a set of interactive events generated by the natural ecosystem both in terms of individual events and continuous processes. It is necessary to clearly separate events that arise in weather situations, i.e., homogeneous and individual events in the natural ecosystem, from the climate, which is expressed by long-term parameters.

In this view, weather changes are clearly due to climate change. Both such activities have a direct impact on business, the economy and society. The climatic conditions in the socio-economic ecosystem are both the result and the cause of human-generated factors as well as the ecosystem itself. For many years, there has been a discussion about what is the result of human actions and what is a natural process from the revolutionary or evolutionary perspective of the ecosystem itself. Economic activity is conducted in such a space. This is particularly important for organizations that conduct their business processes in different locations in different regions and countries. In each of these places, the natural ecosystem is different. As such, other hazards and risks are generated, and other climate opportunities also arise.

The strategic aspect of climate conditions is now becoming one of the core attributes of professional management by organizations that wish to create and understand their own business space and influence their decision-making processes as much as possible. Core technological, ecological, social and political factors emerge against the background of such a system. The pursuit of climate value differs in different regions of the world. It stems from predefined and recognized social and ecological values (Gregori and Holzmann 2020; Dohrmann et al. 2015). This is one of the main barriers to creating a positive climate and climate change around the world today. The European Union is aware of climate risks and is understandably sensitive to them, and in this sense is trying to neutralize their adverse impact, inter alia, through the principles of digital and climate transformation. The key strategic dilemma that arises, among other things, is the answer to the following question: Will the combined implementation of the principles of the digital climate transformation in the long term not destroy the climate and the world as we know it today? Will linking digital solutions to the climate in the long term not generate more opportunities than benefits? An example of today's understanding of such problems are, for example, organized courses to counter addiction to new technologies, including mobile phones, for example through forest treatment therapies. The second example is the fight against the digital exclusion of older people and those who cannot afford access to mobile phones, computers and other elements of modern computer technologies. Hence, the understanding of the core factors of climate impact in the context of activities neutralizing climate effects in the ecosystems of human existence, including the economic one, is becoming a fundamental issue. The table below presents the summary of a set of factors of climate impact on ecosystems of human existence, including the economic ecosystem, against the background of activities neutralizing climate effects in said ecosystems. It also refers to current socio-economic trends affecting the climate and to selected types of climate impact and the recipients thereof.

The Table 16 shows the high level of complexity of issues related to changes and climate anomalies and dysfunctions; as a result, an individualized, related set of climate hazards emerges, as well as a number of strategic and tactical–operational problems concerning the climate and its changes, and as a result, a separate system of an appropriate cause-and-effect nature that has a direct impact on the organizational, management, technological, diagnostic and economic processes in the dynamic management of various types of organizations (Jebeile and Roussos 2023).

To a certain extent and in a certain logic of conduct, a specific state arises as a macroeconomic climate hub, the output of which is a specific system of complex strategic information, which can be used well or improperly at the level of individualized, separately dedicated conditions targeted at individual sectors of the economy in terms of climate change.

6.3 Sectoral determinants of the economy—the climate dimension examined from the perspective of the concept of business models

The sector and/or industry is a very important link in a complex economy. Anyone who does not understand specific sectoral conditions well and correctly will always make an interpretative error in business terms. Sectoral conditions strongly depend on macroeconomic conditions at the state level and on microeconomic conditions at the level of specific companies operating in the sector. The sector sets detailed technological, organizational and economic criteria and has its own internal, unfortunately consumable, intrinsic potential exploited over time. Anyone who can properly read and respond to changes in sectors has a chance to function in the long term regardless of cyclical conditions. Organizations identify crises or disruptive events by looking at specific aspects of them in different areas of operation (Coccia 2020, pp. 81–96). This also applies to sectoral conditions. The characteristics and parameterization of different sectors of the economy determine their crucial potential and business and social opportunities in the context of climate change. In semantic

Table 16. Selected, core and relevant climate impact factors in the context of activities neutralizing climate impact on the ecosystems of human existence, including the economic ecosystem.

No.	Factors of climate impact on the ecosystems of human existence, including the economic ecosystem	Actions neutralizing climate impact on the ecosystems of human existence, including the economic ecosystem
1.	Increased CO_2 emissions	Reducing carbon dioxide emissions, Maintaining the effect of clouds, Increasing technological geoengineering solutions, Increasing climate protection technologies, Rising climate interventions, Reducing climate failures, Rising recognition technologies, Creating a new food system, Mapping industrial emissions, Increasing efforts to decarbonize transport, Reducing consumerism, Reducing purchasing needs, Reducing carbon footprint, Reducing waste and managing waste efficiently, Destroying the recycling 'myth', Implementing the principles of climate justice, Eliminating nationalism, Balancing military forces, Reducing geopolitical inequalities, Building a global agreement for rapid decarbonization, Increasing climate self-awareness, Building regenerative marine agriculture, Developing "Blue Coal" (Water, Macroalgae), Protecting the coastlines of the seas and oceans, Developing methods for the mathematical forecasting of climate change and disasters, Developing research into the impact of global warming on human productivity, Eliminating "irreversible climate impact", Systemic implementation of economic climate packages, Climate levies, Energy storage
2.	Increased emissions of various types of aerosols	
3.	Increased nitrous oxide emissions	
4.	Increased nitrogen emissions	
5.	Loss of stratospheric ozone	
6.	Increased global mean surface temperature of the Earth	
7.	Increased ocean acidification	
8.	Loss of tropical forests	
9.	Increased land use and land urbanization	
10.	Biosphere degradation on land	
11.	Sea level rise	
12.	Increased average global sea surface and land-based air temperatures	
13.	Increased wind speed	
14.	Changes in global air circulation with changes in ocean currents	
15.	Increased critical elements related to climate destabilization in the terrestrial ecosystem	
16.	Increased frequency of changes in the shape of water, the transformation of ice into liquid, from liquid to vapor, and vice versa	
17.	Increased melting of continental ice sheets	
18.	Increased amount of microplastic	
19.	Reduction in the amount of fresh water that is safe to drink	
20.	The paradox of biomass combustion for energy	
21.	Changes in global fire patterns	
22.	Declining insect population	
23.	Phenological changes as the transformation of the status of plants and animals	
24.	Increased melting of permafrost and breakdown of underwater clusters resulting in destabilization of frozen methane, e.g., in the Arctic	
25.	Increased number of viruses that can cause global pandemics	
26.	Increased heat-induced mortality among people	
27.	Increased human mortality from air pollution	

Table 16 contd. ...

...Table 16 contd.

No.	Factors of climate impact on the ecosystems of human existence, including the economic ecosystem	Actions neutralizing climate impact on the ecosystems of human existence, including the economic ecosystem
28.	Increased vector diseases transmitted between humans and different species of animals and other beings and vice versa	
29.	Increased bacterial resistance to antibiotics	
30.	Increased plagues in agricultural systems	
31.	Increased climate migration, increased number of climate refugees	
32.	The rise of environmental racism	
33.	The fall and elimination of indigenous nations	
34.	Increased number of global inequalities caused by climate change	
35.	Increased number of natural disasters	
36.	Increased number of climate conflicts (the likelihood of different types of interpersonal conflicts increases with rising temperatures)	
37.	The "intentional" emissions error paradox, the "net-zero emissions" paradox	
38.	Insufficiently rapid transition to renewable energy (solar energy, wind energy, green hydrogen, hydropower, nuclear energy, biomass energy, geothermal energy, using and saving less energy, creating more energy efficient goods and buildings, consuming energy where and when it is produced)	
39.	Increased deforestation	
40.	No systemic changes in the structure of food for less caloric products and at the same time no changes in systemic eating habits	
Socio-economic trends affecting the climate		World population growth, Real GDP growth, Direct investment growth, Urban population growth, Primary energy consumption growth, Fertilizer consumption growth, Large dam construction growth, Water consumption growth, Paper production growth, Means of transport growth, Mobile phone subscriber connection growth, International tourism growth
Selected Climate Impact		Greenhouse effect, Global warming, Increased number of droughts and floods, Increased number of hurricanes, Increased number of fires, Increased storms and heavy rainfall, Decreased snowfall, Increased number of storm waves
Recipient of the Climate Impact		The environment as a global entity

Source: Jabłoński and Jabłoński (2024, pp. 48–49)

terms, the term 'climate destabilization and dysfunction' should be used rather than climate change. Unfortunately, the negative climatic impact can also be imagined at a given level of specifically managed organizations operating on a specific market in the economy. Thus, different types of climate-related opportunities and threats are generated in such a specific thinking space. The total sum of the potential opportunities and threats present is influenced by the unique image and shape of a given sector. In the context of these climate opportunities and threats, the designated technological and innovation potential of a given sector is vitally important here. Technological potential is a platform for building products and services and characterizes new and modern mechanisms built in core production and/or service processes. When categorizing individual sectors of the economy, particular attention should be paid to the fact that they can be highly climate-influencing or moderately climate- and climate-change-influencing sectors. In both cases, business models and strategies will then be different and implemented differently. In each of these specific cases, the business models and strategies implemented can be reactive or proactive. In many cases, there is no comprehensible awareness among managers of the climatic impact of technological processes in production and production processes that are particularly important in the context of current sectoral conditions. The creation of climate-positive technologies that are important for modern circularity-based production processes is becoming crucial (Bojovic et al. 2021). These technologies usually have very high innovation potential. Then, there is a transition from linear business models to circular business models (Jabłoński and Jabłoński 2023). The transition to a circular and digital business model also changes the logic of cost management in the company to so-called marginal cost accounting. It should be noted that the climate and ecological impact of technological processes has so far been mainly focused on reducing emissions to air, emissions to water, emissions to soil, resource use and waste management. However, this is indeterminate, because the so-called post-production climate effect measured by the direct impact of production processes on the climate of a given, specific company is important. Therefore, it is essential to link individualized, specifically defined sectoral conditions, which may constitute only a canvas as a specific platform for designing and implementing business models and strategies in the context of emerging climate change. Linking the individual awareness and climate sensitivity of managers together with the institutional building of climate culture can only give a clear final result.

This link should be mutually coherent and focused on the principles of such design of business models and such definition of long-term goals in the organization that will shape the mechanisms of effective and efficient operation of the organization and support the climate policies of individual countries (Panenko et al. 2021).

The climate problems presented are complex and often ambiguous. Solutions to these problems can therefore be implemented in different ways and from different perspectives. This is primarily related to the resulting costs associated with them and the specific and properly applied monitoring and measurement of the effectiveness of the economic results achieved and other benefits created. Description of the main climate challenges for sectors of the economy particularly vulnerable to climate change shows Table 17.

Table 17. Description of the main climate challenges for sectors of the economy particularly vulnerable to climate change.

No.	Sectors of the economy particularly vulnerable to climate change	Description of main climate problems
1.	Agriculture	Agriculture has a significant impact on climate change, being one of the main greenhouse gas (GHG)-emitting sectors. The main climate problems that have a strong impact on agriculture include weather unpredictability, air temperature rise, sea level rise, and droughts.
2.	Energy Sector	Conventional power plants, such as coal or nuclear power plants, are prone to failures as a result of extreme weather conditions. The increased temperature results in an increased use of air conditioning and other refrigeration equipment, which in turn increases the demand for electricity. Solar, wind, hybrid and biomass energy are more environmentally friendly and less susceptible to the negative effects of extreme weather events. Power plants, transmission lines and energy storage systems need to be more resilient to extreme weather events.
3.	Tourism	Climate change manifesting itself in the increasingly frequent occurrence of non-extreme weather events, problems with excess or lack of water, fires, movement of plant and animal species caused by changes in weather conditions and other phenomena related to the climate.

Source: Own study.

6.4 Designing and implementing business models and strategies in the context of climate change—sectoral determinants

The business model, as a core and recognized ontological entity in strategic management, is a platform for the market success and failure of companies. Today it is already obvious and extremely visible that companies do not compete only by means of products and services, but above all business models and their unique configuration. A market-winning business model is one that follows the logic of sector determinants and is understood by managers, employees, suppliers and customers. Interfaces between internal and external components of the business model in its canvas shape the mutual relationships between entities involved in the dynamics of a given component of the business model. The design and structure of a business model is determined by the specific and individual characteristics and attributes of the specific sector in which the model is used by the organization. This also applies to strategies, the attributes of which defined in the mesoenvironment shape a specific dimension of the set and implemented long-term goals of the organization.

Each canvas of the business model should be associated with attributes that are built up and describe a given sector of the economy. There are many attributes describing a given sector of the economy. On the one hand, these are specific groups of general issues and, on the other hand, they are very detailed, precise and specific factors for a given sector. The group of these issues determining the existential nature and dimension of a given sector may include its competitiveness, attractiveness, stability, sector life cycle, economic potential, technological potential and social potential.

The attributes of the economic sector identified above should clearly express its key potential, which should be taken advantage of appropriately by the companies embedded in it. It should be noted that the defined attributes should be oriented towards social and economic expectations with a proper understanding of climate problems. This should not only be considered in the category of global warming, but above all in relation to the rapid dysfunctions and destruction of the socio-economic ecosystem caused by climate change anomalies. Then, the competitiveness, attractiveness and stability of the sector will be understood in a completely different way, the full life cycle of a given sector will be shaped differently, and its technological and social economic potential will evolve. Such a construction and market structure of the sector can be the main determinant for building and implementing the appropriate boundary conditions necessary to design business models and strategies that will give the greatest final value. They and a substantive platform underline the construction of the business models and strategies of companies. Any mismatch between the structure of attributes of the economic sector and the configuration of the business model and a dedicated stream of long-term goals of the organization will generate negative value rather than added value. The gap between sectoral conditions and the construction of the business model and strategy will turn into unused capital, which may consequently lead to the destruction of the value of a specific organization embedded in a given sector. Organizations managed wisely by conscious and sensitive managers will not make this mistake. It is therefore important to know the sector and the mechanisms for effective and efficient construction of business models and strategies. In configuration terms, the business model is an aggregated set of components linked together by relevant interfaces, constituting a system with a designated function in the goal (Kulins et al. 2016). This goal is usually to achieve acceptable effectiveness and efficiency, measured by the attributes of the constantly increasing long-term value of the organization. The business model converts payments received into profits (Teece 2010, pp. 172–194). The business model then generates a comprehensible value proposition properly received for the customer and expected by him, which he is ready to pay for. Appropriate income generation logic by the organization shapes, from the same perspective, the resultant, economic and social final profit of the company. In the case of expanding the group of components of the business model and the stream of long-term goals of the organization with climate aspects, the expected climate profit and yield by many stakeholder groups also arises. The climate profit will be the result of an appropriate ratio of revenues and costs of investing in climate aspects, and the yield will be left as a potential resource in various types of actions aimed at climate change.

6.5 Business models and climate change strategies focused on the resulting negative, neutral, and positive dimensions

The business models of companies create huge potential for market impact and are a tool for achieving competitive advantages on the market. As a competitive factor, they are imitated. A good, proper business model is a model that is very difficult for other entities to reproduce and copy because it is resistant to counterfeiting and duplication. However, this is not so easy, as it requires the construction of safe systems to protect against such practices carried out by competitors. A Business model is a system that solves the problem of identifying who is (or are) the customer(s), engaging with their needs, delivering satisfaction, and monetizing the value (Baden-Fuller and Haefliger 2013, pp. 419–426). The open business model specifically describes value creation and capture by “systematically collaborating with outside partners” (Osterwalder and Pigneur 2010). The business model concept has been used to explain how companies create and derive value from technological innovation (Chesbrough and Rosenbloom 2002, pp. 529–555). Changes in the structure and relationships of the business model can be divided into three broad categories, namely automation, expansion and transformation (AET) (Massa and Tucci 2012). The construction of business models and strategies is not a simple and unambiguous issue. In addition to the applied attributes of sectoral conditions, it is important to properly build the configuration of intangible resources constituting the canvas of the business model and a strategy map with strategic goals embedded therein. What is also important is the creative action of managers based on micro-objectives, micro-actions and making various types of micro-corrections in every area of the organization’s existence relevant to a given situation. In the process of designing business models, attention should be paid to the appropriate selection of their components. The component structure of a business model is usually contained in the linkage of its general components with specific components dedicated to the corresponding attribute expressed in that business model. For instance, in a given business model, the following attributes can be expressed: scalability, consistency, durability, creativity and resilience. If we refer to climate change in the business model, the statement that a given business model is rich in intelligent climate values, poor in these climate values or climate-neutral can be used. These value drivers are the components built into the business model (Spieth et al. 2018). In the case of a business model rich in climate values, we are dealing with a business model that positively affects the climate. In the case of a business model poor in climate values, we can call it a business model that negatively affects the climate. The third type of business model is a climate-neutral business model. At this point, it is worth noting that business models and strategies rich in intelligent climate values can be additionally equipped with specific predictive mechanisms used, on the one hand, for multidimensional measurement and monitoring of technological and production processes and, on the other hand, building a tight information system that will meet the appropriate criteria for dedicated and applied climate services.

Then, the value proposition for the customer will be a product that positively affects the climate. The income generation logic in its processes will then fulfill the designated requirements of the circular economy and a low- or zero-emission economy, and the system of measuring and monitoring processes in the organization will then be based on mathematical and statistical models equipped with artificial intelligence solutions (Di Vaio et al. 2020, pp. 283–314; Francisco 2023), which will be determined in the relevant climate services also provided by organizations. Therefore, it is advisable to link the proper construction of business models with technological and production processes and strategic information resulting from the prediction of climate change based on climate services. Climate efficiency can be achieved by implementing circularity principles that generate waste-free production. At the company level, this may also include the principles of explicit climate risk, i.e., one that can be imagined and that can assess the degree of occurrence of potential climate catastrophes within the places of business existence of the organizations concerned. These types of business models are described and interpreted below.

6.6 Climate-negative business models and strategies generating a negative climate balance

The adverse impact of the economy and business in the absence of appropriate processes to inhibit current emissions into the environment of specific companies is known, obvious and currently strongly condemned through dedicated important formal and legal solutions embedded in the regulatory system of the countries concerned. The absence of specific and controlled point and linear emitters on the negative impact on the natural ecosystem always causes clear and visible ecological and climate problems. Therefore, the elimination of such activities should already be generated at an early stage of the development of the concept of business models of companies in the so-called bottom-up approach, which is of an embryonic nature. One should be aware that the rate of consumption of the natural resources concerned is unsustainable, and therefore the development of new (economic) systems with an emphasis on sustainable practices, technologies and processes is necessary (Bocken et al. 2014, pp. 42–56). Therefore, it is necessary to decisively eliminate or reduce current and emerging business models that change the climate negatively and replace them with positive ones. A business model that changes the climate negatively can be interpreted as a specific value proposition for the customer in the form of products or services that affect him negatively. Income generation logic in the form of the use of specific technological and production processes will then generate a visible adverse impact on the environment and the climate in the form of, inter alia, harmful and excessive emissions of illicit agents, raw materials and components into the air, water and soil. It will also significantly over-exploit resources, including natural resources, and the waste management system will be operated in an inappropriate and often uncontrolled way at all stages of production and in post-production processes. Cost optimization will take place without specific waste-free or low-carbon solutions. In such a business model and given its logic and methodology of use and operation, the circular production management model will not be used in any way, but only a classic linear one, generating a wide waste stream. There will also be no set of ecological and climate indicators with a defined point of achievement, which will be systematically monitored and will form the basis for effective and efficient management of climate risks at the level of a given organization. Solutions based on recycling and the reuse of raw materials in circular processes will also not apply and appropriate climate and ecological competences and skills will not be built in a given organization. The concept of "cradle to cradle" will not apply here. Such a company will not understand or, for example, strive to reduce various types of emissions in the processes of managing the entire supply chain for customers. With regard to the strategies used that negatively affect the climate, this will be a set of long-term goals of the organization which in no way will be aimed at neutralizing or eliminating the negative impact on the environment and the climate. These strategies will be aimed at fully maximizing profit for shareholders according to the classic approach of technocratic capitalism. Of course, such strategies are short-sighted and ineffective in the long term. In this type of strategy, purely economic goals will usually appear, sometimes even economically aggressive ones, aimed at the lack of balance among the expectations of other groups of stakeholders interested, among other things, in positive climate change, and ruthless greed and the desire for money will obscure any other goals of this organization. It should be noted that at the level of a comprehensive assessment of the impact of a particular company on the natural ecosystem and climate, it is not only the mechanism of CO_2 production that is decisive. This is a much broader dimension, due also to the fact that the company, as a social, technological and economic entity, affects numerous factors of human existence. Therefore, a holistic, comprehensive view of the organization in terms of its negative impact on the environment and climate is advisable in a systemic and process-based way, based on facts and evidence rather than opinions and other types of digressions. In this particular case, only facts resulting from reliable scientific research should be taken into account. The principles of evidence management in management and quality sciences can be used for this purpose.

The (Table 18) description of the attributes of a business model which negatively changes the climate presented here defines, to some extent, even an archaic company, i.e., one that has not kept up with technological changes within the developed and applied ecological and climate criteria. It can therefore be assumed that this organization has deliberately or unknowingly fallen into its own trap of strategic and

Table 18. Description of attributes of a business model which negatively changes the climate.

No.	Attributes of a business model which negatively changes the climate
1.	Obsolete and inefficient production processes
2.	A large amount of waste and high emissions
3.	Linear production–sales–disposal at the landfill
4.	Underutilized capacity, unused assets and resources
5.	Unused byproducts of production
6.	Extraction and use of fossil fuels and non-renewable or finite resources
7.	Use of polluting, unnatural mining and production processes (Bocken and Short 2021).

Source: Own study.

technological jamming. The degree of opportunity for the strategic renewal of such an organization from the perspective of a dynamic and rapidly advancing digital and climate transformation is almost zero.

6.7 Climate-neutral business models and zero climate balance strategies

Being neutral is not exactly the climate impact we would really want. However, it is always necessary to aim higher, especially in the area of improving the quality of human existence at the level of specific local and global communities. Climate neutrality is a very important concept that defines the balance between greenhouse gases and their storage or absorption by water reservoirs, forests or soils. The replacement of coal-fired generation sources with renewable energy sources has created a new energy sector. Energy transformation has changed the structure of the energy sector (Giehl et al. 2020). It often happens that the configuration of a specific business model built from the value proposition expected for the customer and the dedicated income generation logic in the climate and natural environment criterion is neutral. This is due to the specific sectoral circumstances in which the company functions and operates on the market. It also depends on the type of product or service provided as well as the specifically developed and required boundary conditions for the conduct of production and post-production processes, including designated mechanisms for delivering the finished product to the customer, usually at a specific place of its creation. A situation when the sector and its specific organizational and technological parameters do not have a significant impact on the environment may also occur. This will result from this unique specificity of a given sector and/or industry. The specificities of this sector will obviously be clearly and closely linked to the products or services created in the company. In the context of an environmentally neutral strategy, the long-term goals of the organization will therefore appear, where the contribution and incentive will not be related to environmental factors and climate change. Then, the dynamic balance between the individual long-term goals of this organization will not take into account the activity in the area of developing appropriate pro-ecological and pro-climate behaviors. The neutrality of the business model and strategy presented here may be the result of natural parameters of the functioning of a given organization in a given sector. It may also be a result of the decision-making processes of managers resulting from the adopted intentional nature of their understanding of running a business. Such logic and interpretation leads to the fact that the effect of a climate-neutral business model is the climate neutrality of the company implementing this business model. Description of the attributes of a climate-neutral business model to introduce Table 19.

The specified attributes of a climate-neutral business model are adopted in a progressive approach to climate change at a minimum level, the so-called hygienic level, but do not often go beyond the relevant regulatory and/or formal legal requirements. They simply provide the organization with peace, but they are not a particular strategic differentiator or a leading platform for generating a unique, climate-innovative competitive, technological, ecological or climate advantage.

Table 19. Description of the attributes of a climate-neutral business model.

No.	Attributes of a climate-neutral business model
1.	Systemic reduction of the installed capacity of conventional power and heat generation sources with high emission indicators in the medium term and a change in the fuel mix through the increased share of zero- and low-emission sources.
2.	Implementing climate innovation at a minimum level.
3.	Applying circular economy solutions to the legally required extent.
4.	Using e-products as an alternative to analog products.
5.	Implementing a comprehensive cybersecurity management system in sufficient terms.
6.	Reducing the carbon footprint in the value chain, where possible and cost-effective.

Source: Own study.

Additionally, they show the clearly untapped intellectual potential of organizations in a space that can have a strong application for significant climate innovations.

6.8 Business models which positively change the climate and strategies which generate a positive climate balance

The positive impact on climate change and the search for a wide stream of innovative solutions should be a strong and deliberate action of wise, creative and climate-creative visionaries of the theory and practice of business and the economy. They should be supported in various possible ways to ensure that people with such a dynamic innovation profile unleash their huge potential for intellectual opportunities for the climate. Both formal and informal instruments should be developed for these processes, aiming to build suitable incentive systems for innovators and creators of positive climate change. A separate and highly specific group of business models and strategies will be business models and strategies that positively change the climate. In constructing such business models and strategies, the polarization of the strategic thinking of different groups of decision-makers towards possible climate opportunities rather than threats and risks must be resolutely and clearly reversed. This refers to a good understanding of these strategic changes and transforming them from emerging risks and threats into appropriate market opportunities to create an innovative economy that positively affects the climate. What is needed here is keen anticipation by the organization's data managers, i.e., their strategic ability to predict possible disturbing market phenomena and quickly convert them into positive business results. This can be achieved by creating climate-positive products and services through dedicated climate-changing business models and strategies. Then, a value proposition for the climate-sensitive customer is climate-positive products and services, including climate services (Hewitt and Stone 2021; Bessembinder et al. 2019). In terms of a specific income generation logic, it is the use of automation, robotics, technical diagnostics, mechatronics, cybernetics and artificial intelligence in effective production and service management based on the assumptions of the circular economy and in the area of economics and finance based on the principles of marginal costs. The catalyst and accelerator of these dynamic changes should be, among other things, the effectively and efficiently implemented principles of digital transformation. It should be noted that the principles of the circular economy together with the principles of digital transformation, i.e., the so-called double transformation, have a strong impact on the economic mechanisms of various types of organizations (Chiappetta Jabbour et al. 2020). At the same time, the cost of climate change will become an economic and social gain. In terms of climate-changing business models and strategies, an action taken by every competent manager should therefore be to understand the mechanisms of climate strategy and policy in Europe and other continents. For example, if CO2 emissions in Europe will have been almost completely eliminated from domestic and technological processes by 2050, innovative technological solutions embedded in business models should be built based on these assumptions. Bioenergy with carbon dioxide capture and storage is one example of existing CO_2 removal techniques (CDR). CDR strategies are defined as "efforts to remove CO_2 from the atmosphere" (National Research Council 2015). This should be achieved, inter alia, through the stream of long-term climate objectives of the company concerned. In this case, a functional climate strategy of the company resulting from its broadly understood climate responsibility should be embedded in the management mechanisms.

The (Table 20) description of attributes of a business model which positively changes the climate clearly shows the scheme of understanding and perceiving these attributes and defines the intellectual and social capital of the solutions set out here clearly and transparently. The number of these strategic changes within the designated positive climate adaptation creates a new shape and image of a climate-responsible company in a specific space of business and social conditions which are definitely worthy of a wide and multidimensional analysis and the strategic synthesis and continuous improvement of climate-conscious and sensitive organizations.

Table 20. Description of the attributes of a business model which positively changes the climate.

No.	Attributes of a business model which positively changes the climate
1.	Implementing full, comprehensive principles of the circular economy in production processes in company logistics
2.	Implementing the principles of a zero-emission and zero-waste economy throughout the company's value chain
3.	A comprehensive link between the principles of digital and climate transformation
4.	Implementing comprehensive principles of automation, robotics, mechatronics, technical and measurement diagnostics and artificial intelligence in all core links and areas of the company's operation
5.	Building mechanisms of the offered proposal of intelligent climate value in the business model and climate strategy
6.	Creating climate-innovative services and products as a result of a climate-positive business model

Source: Own study.

6.9 Conclusions

The logic of understanding climate change in this way should be very intensively designed, implemented and substantively cascaded from macroeconomic conditions to mesoeconomic, i.e., sectoral, to microeconomic, i.e., at the level of specific companies. In addition, the dimension of examining and applying such business models, together with their specific value proposition and the logic of generated revenues, should include a multi-criteria and multi-faceted assessment thereof against the background of positive, neutral and negative climate changes occurring in reality. Only then can one be certain that the mechanisms of strategic management for the climate and its dynamic changes are properly shaped. Negative climate effects can then be stopped through political, social and economic decisions.

References

Baden-Fuller, Ch. and Haefliger, S. (2013). *Business Models and Technological Innovation*, Long Range Planning 46: 419–426, http://dx.doi.org/10.1016/j.lrp.2013.08.023.

Bessembinder, J. et al. (2019). *Need for a common typology of climate services*, Climate Services, https://doi.org/10.1016/j.cliser.2019.100135.

Bocken, N.M.P. and Short, S.W. (2021). *Unsustainable business models – Recognising and resolving institutionalised social and environmental harm*. Journal of Cleaner Production, 127828. DOI: https://doi.org/10.1016/j.jclepro.2021.127828.

Bocken, N.M.P., Short, S.W., Rana, P. and Evans, S. (2014). *A literature and practice review to develop sustainable business model archetypes*. J. Clean. Prod. 65: 42–56.

Bojovic, D., St. Clair, A.L., Christel, I., Terrado, M., Stanzel, P., Gonzalez, P. and Palin, E.J. (2021). *Engagement, involvement and empowerment: Three realms of a coproduction framework for climate services*, Global Environmental Change 68: 102271, https://doi.org/10.1016/j.gloenvcha.2021.102271.

Chiappetta Jabbour, Ch. J., Seuring, S., Lopes de Sousa Jabbour, A.B., Jugend, D., De Camargo Fiorini, P., Latan, H. and Colucci Izeppi, W. (2020). *Stakeholders, innovative business models for the circular economy and sustainable performance of firms in an emerging economy facing institutional voids*, Journal of Environmental Management 264: 110416, https://doi.org/10.1016/j.jenvman.2020.110416.

Chesbrough, H. and Rosenbloom, R. S. (2002). *The role of the business model in capturing value from innovation: Evidence from Xerox Corporation's technology spin-off companies*. Industrial and Corporate Change, 11: 529–555.

Coccia, M. (2020). *Critical decision in crisis management: Rational strategies of decision making*. Journal of Economics Library, 7(2): 81–96.

Di Vaio, A., Palladino, R., Hassan, R. and Escobar, O. (2020). *Artificial intelligence and business models in the sustainable development goals perspective: A systematic literature review*, Journal of Business Research 121: 283–314, https://doi.org/10.1016/j.jbusres.2020.08.019.

Dohrmann, S., Raith, M. and Siebold, N. (2015). *Monetizing Social Value Creation – A Business Model Approach*, Entrep. Res. J.; 5(2): 127–154, DOI 10.1515/erj-2013-0074.

Dolsak, N. and Prakash, A. (2018). *The Politics of Climate Change Adaptation, Annu. Rev. Environ. Resour.* 43: 317–41.

Fischer, M., Imgrund, F., Janiesch, Ch. and Winkelmann, A. (2020). Strategy archetypes for digital transformation: Defining meta objectives using business process management, Information & Management 57: 103262, https://doi.org/10.1016/j.im.2019.103262.

Francisco, M. (2023). *Artificial intelligence for environmental security: national, international, human and ecological perspectives*, Current Opinion in Environmental Sustainability, 61: 101250, https://doi.org/10.1016/j.cosust.2022.101250.

Giehl, J., Göcke, H., Grosse, B., Kochems, J. and Müller-Kirchenbauer, J. (2020). *Survey and Classification of Business Models for the Energy Transformation. Energies*, 13(11), article 2981. https://doi.org/10.3390/en1311981.

Gregori, P. and Holzmann, P. (2020). *Digital sustainable entrepreneurship: A business model perspective on embedding digital technologies for social and environmental value creation*, Journal of Cleaner Production 272: 122817, https://doi.org/10.1016/j.jclepro.2020.122817.

Hewitt, C. D., Guglielmo, F., Joussaume, S., Bessembinder, J., Christel, I., Doblas-Reyes, F.J., Djurdjevic, V., Garrett, N., Kjellström, E., Krzic, A., Máñez Costa, M. and St. Clair, A.L. (2021). *Recommendations for Future Research Priorities for Climate Modeling and Climate Services*, American Meteorological Society, https://doi.org/10.1175/BAMS-D-20-0103.1.

Hewitt, Ch.D. and Stone, R. (2021). *Climate services for managing societal risks and opportunities*, Climate Services 23: 100240, https://doi.org/10.1016/j.cliser.2021.100240.

Jabłoński, A. (2013a). *Modele zrównoważonego biznesu w budowie długoterminowej wartości przedsiębiorstw z uwzględnieniem ich społecznej odpowiedzialności* [Sustainable business models in building the long-term value of companies with consideration of their social responsibility], Difin, Warszawa.

Jabłoński, A. (2013b). Zielony wymiar budowy modeli biznesu i strategii przedsiębiorstw [The green dimension of building business models and company strategies], [w:] Urszula Wąsikiewicz-Rusnak (Ed.) (collective work), Współczesne problemy rozwoju przedsiębiorstwa w koncepcji rozwoju zrównoważonego [Contemporary problems of company development in the concept of sustainable development], WSB Dąbrowa Górnicza.

Jabłoński, A. (2015), Design and Operationalization of Technological Business Models, Acta Universitatis Agriculturae et Silviculturae Mendelianae Brunensis, Vol. 63, No. 3.

Jabłoński, A. (2023). *Usługi klimatyczne w kształtowaniu modeli biznesu przedsiębiorstw* [Climate services in shaping business models of companies], Przegląd Organizacji 4/2023.

Jabłoński, A. and Jabłoński, M. (2023). *Determinanty transformacji linearnych modeli biznesu w stronę orientacji cyrkularnej* [Determinants of transformation of linear business models towards circular orientation], Prace Naukowe Uniwersytetu Ekonomicznego we Wrocławiu, Research Papers of Wroclaw University of Economics and Business, Vol. 67, No. 4, ISSN 2392-0041.

Jabłoński, M. and Jabłoński, A. (2024). *Projektowanie i wdrażanie modeli biznesu i strategii zmieniających klimat w kryterium zarządzania ryzykiem klimatycznym. Uwarunkowania sektorowe na przykładzie sektora transportu kolejowego* [Designing and implementing business models and strategies that change the climate in a criterion for climate risk management. Sectoral conditions –the case study of the railway transport sector], Zeszyty Naukowe Akademii Górnośląskiej [Research Bulletin of Katowice Business University], No 4/2024 (16), pp.48-49.

Jebeile, J. and Roussos, J. (2023). *Usability of climate information: Toward a new scientific framework*, WIREs Clim Change; 14: e833. wires.wiley.com/climatechange, https://doi.org/10.1002/wcc.833.

Kohtamäki, M., Parida, V. Oghazi, P., Gebauer, H. and Baines, T. (2019). *Digital servitization business models in ecosystems: A theory of the firm*, Journal of Business Research 104: 380–392, https://doi.org/10.1016/j.jbusres.2019.06.027.

Kulins, CH., Leonardy, H. and Weber, Ch. (2016). *A configurational approach in business model design*, Journal of Business Research 69: 1437–1441, http://dx.doi.org/10.1016/j.jbusres.2015.10.121.

Larosa, F. and Mysiak, J. (2020). *Business models for climate services: An analysis*, Climate Services 17, 100111, https://doi.org/10.1016/j.cliser.2019.100111.

Massa, L. and Tucci, C. (2012). *Business model innovation. In*: Dodgson, M., Gann, D. and Phillips, N. (Eds.). *The Oxford Handbook of Innovation Management*. Oxford University Press, London.

Mentink, B. (2014). *Circular Business Model Innovation A process framework and a tool for business model innovation in a circular economy*. MSc thesis Industrial Ecology Master Theis. Delft University of Technology in Delft (Holland) and Leiden University in Leiden (Holland).

National Research Council (U.S.). (2015). *Climate Intervention: Carbon Dioxide Removal and Reliable Sequestration*, National Academies Press, Washington, D.C.

Orlove, B. (2022). *The Concept of Adaptation, Annu. Rev. Environ. Resour*. 47: 535–81.

Osterwalder, A. and Pigneur, Y. (2010). *In*: Clark, T. (Ed.). *Business model generation: A handbook for visionaries, game changers, and challengers*. Wiley.

Panenko, A., George, E. and Lutoff, C. (2021). *Towards the development of climate adaptation knowledge-action systems in the European Union: An institutional approach to climate service analysis*, Climate Services 24: 100265, https://doi.org/10.1016/j.cliser.2021.100265.

Paul, A., Lang, J.W.B. and Baumgartner, R.J. (2017). *A multilevel approach for assessing business strategies on climate change*, Journal of Cleaner Production 160: 50–70, http://dx.doi.org/10.1016/j.jclepro.2017.04.030.

Rochlin, C. (2021). *Climate change and the progressive business model*, The Electricity Journal, Vol. 34, No. 4, art. 106927.

Rosenstock, T.S., Lubberink, R., Gondwe, S., Manyise, T. and Dentoni, D. (2020). *Inclusive and adaptive business models for climate-smart value creation*, Current Opinion in Environmental Sustainability, 42: 76–81.

Spieth, P., Schneider, S., Clauß, T. and Eichenberg, D. (2018). *Value drivers of social businesses: A business model perspective*, Long Range Planning https://doi.org/10.1016/j.lrp.2018.04.004.

Teece, D.J. (2010), *Business Models, Business Strategy and Innovation*, Long Range Planning 43: 172–194, doi:10.1016/j.lrp.2009.07.003.

Yunus, M., Moingeon, B. and Lehmann-Ortega, L. (2010). Building social business models: lessons from the Grameen experience, "Long-range Planning", 43(2/3): 308–325.

7

Climate Responsibility in the Technological Management of Companies and their Business Models (Digital Transformation, Automation, Robotics, Diagnostics, Artificial Intelligence)

7.1 Introduction

The aim of this chapter is to present the core relationships between digital transformation and climate change processes in the criterion of managerial decision-making. The scope of the chapter includes selected relationships that have a decisive impact on the creation of innovative business models and strategies by managers conducting business which is strongly dependent on the shape and image of the external and internal ecosystems of organizations striving to achieve above-average market results. The subject of the chapter concerns the mechanisms of shaping strategic, tactical and operational changes in the organization towards a specific adaptation to current technological, organizational, management and operational conditions determining the continuity of the organization's operation alongside the ability to dynamically develop them.

7.2 Climate responsibility in the strategic digital management of companies

Technology is changing the entire world exponentially, and this is not just a slogan. Anyone who does not notice this makes a theoretical error, which can quickly turn into a practical error. This is already clearly visible in the economy and business, and in the social field it has also become a visible negative factor in terms of the

DOI: 10.1201/9781003474401-7

technological or digital exclusion of an individual. The dynamics of today's difficult world create new needs and expectations in terms of the development of the economy, social conditions and new needs of various groups and entities. The fulfillment of these needs requires a radically different view of the reality around us than before. The conditions of today's world are diametrically different and more complicated, so the solutions used today should also be different. It is therefore important to deepen the analysis and synthesis of the social and economic ecosystem at the level of the whole world and local communities. This ecosystem and its further evolution is ambiguous and very difficult to understand. It requires very broad and multidimensional knowledge to make the right conclusions and strategic recommendations for various areas of the functioning of these ecosystems. From this perspective of today's social and market situation, a whole range of crises emerge, which add to each other and create various types of problems, the solutions to which must be systemic. Chaos cannot be eliminated by chaos or replaced by other forms of chaos. Moreover, it is crucial to understand the technological and digital changes in today's world, especially in the context of military, health, economic and climate threats. In such a space of change, there is a very strong relationship between the digital transformation and dynamically advancing climate change, the consequences of which are already manifesting themselves in all fields of human life. In the era of strong technological development, the effective management of organizations must be focused on the technological component in managerial decision-making processes in various fields of organizational operation and their key ontological entities. Automation, robotics, diagnostics, and artificial intelligence have incredible intellectual potential that can de facto change the logic of the currently functioning mechanisms of industrialization of the economy. The stream of generated existing and feasible technological solutions at the company level already has a exponential dimension. While searching for breakthrough innovations, companies undertake highly sophisticated experiments to achieve the results they project.

The technological triad of the current economy in developing the principles of climate responsibility is presented in the figure below.

The technological triad presented in the Figure 8, reinforced with artificial intelligence solutions, should be implemented in a climate-responsible manner. For this to happen, it is important to understand the future consequences of technological innovation, which are not simple and unambiguously obvious. Technological innovation geared towards a positive impact on climate change should become a

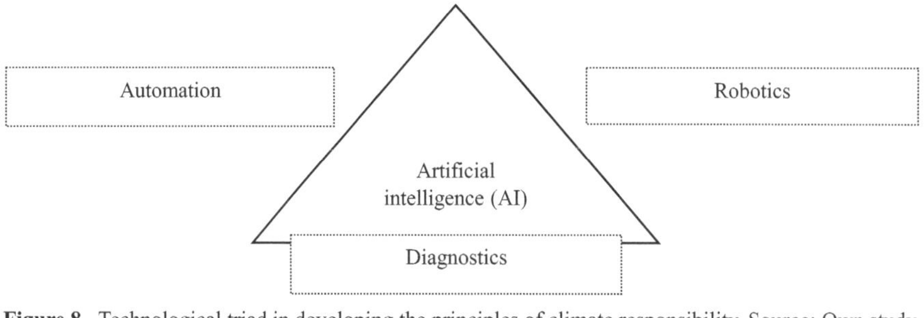

Figure 8. Technological triad in developing the principles of climate responsibility. Source: Own study.

powerful enabler of the new creative thinking space for innovators and business visionaries.

The creation and dynamic development of the digital world is already a fact. Human beings exist on the edge of the real and digital world. In order to properly integrate these two worlds, the ethics and morality of the creators and the responsibility and honesty of users are required. In the current dynamic process of combined digital and climate transformation, climate responsibility takes on a different meaning and image. Entrepreneurs, who are also aware of their social mission, are becoming increasingly aware of climate threats, not least because these threats may also affect the company's infrastructure, leading to its destruction and, consequently, to the possible bankruptcy of the organization. This is especially the case in the parts of the world most directly exposed to anomalies and climate disasters. Moreover, the digital world generates both opportunities and threats. Today's environment is an agile and flexible one where identifying business opportunities and creating business models are required in a very limited time to turn technological innovations into realistic business models. This environment is particularly important if companies are to face up to today's business reality (Xu 2018). In addition, the development of digitalization technologies such as Big Data and Data Science has made business models and processes change enormously, and new business models are emerging like never before, which generates new levels of corporate responsibility (Härting et al. 2018, pp. 1495–1506). Artificial intelligence is a very important tool for combating climate change. For example, it is implemented through a chatbot system that uses AI technology to send messages, e.g., about the rising level of rivers and lakes, landslides, droughts or possible earthquakes. Digital Deloitte defines three main types of chatbots: informational, transactional and advisory. Information chatbots are the simplest and most popular type available, conveying general information. Transactional chatbots allow users to perform a specific action. Advisory chatbots (virtual advisors) are the most advanced form of software, able to learn from interactions with users and thus improve the quality of the next conversation (Digital Deloitte 2017).

From this perspective, it is worth noting that digital and non-digital ventures follow different patterns in the development and maturation of the business model in the early stages of the company's life cycle to achieve the operational ability to generate revenues. The results show that different strategies and support schemes in digital and non-digital ventures are relevant from a longitudinal perspective. In the case of a digital venture, this means in particular that support must initially focus on developing transactions with customers before seeking investment. In contrast, non-digital ventures require prior investment to build capital-intensive assets to create value (König 2016). This applies especially to understanding and building climate values in various types of organizations. Therefore, the principles of the digital green deal should be implemented very wisely. It may be interpreted differently, but it seems that the main challenge in this case is to shape the communication message expressed by real evidence on this topic in the area of revenues and costs, revenues and expenses, and costs and benefits at the level of companies and countries associated with the climate impact.

7.3 Eco-marketing or greenwashing? In the space of climate responsibility

Is it possible to create responsible environmental and climate marketing? This may be so if well understood and used, but it is not easy and unambiguous in terms of social and business perception. Such a challenge should be a priority for the responsible managers of responsible organizations. In business, following ethical criteria in the values adopted for a given place and space by the majority of society should not only be a priority but, above all, mandatory behavior, while at the same time being noticed and appreciated by the community.

Green marketing can be defined as the use of marketing tools to facilitate trade that meets both organizational and individual goals, while protecting and preserving the physical environment (Lam and Li 2019). Eco-marketing, also called green marketing, combines advertising activities with the promotion of pro-ecological attitudes and behaviors. This is the sum of positive and ethical actions for the protection of the environment, which can include actions of various types, e.g., aimed at reducing excessive consumerism, promoting recycling solutions or counteracting environmental destruction. Green marketing is also seen as an organization's commitment to strategic, tactical, and internal activities and processes, the holistic goal of which is to create, communicate, and deliver products with minimal environmental impact (Papadas et al. 2017, pp. 73–81).

The term "greenwashing" was first used in 1986 by environmentalist Jay Westervelt, who published an essay on the hotel industry about its practices promoting the reuse of towels (Walker and Wan 2011, pp. 227–242). Greenwashing is defined as "poor environmental performance and positive communication on environmental performance" (Delmas and Burbano 2011, pp. 64–87). It is also treated as "the act of misleading consumers regarding the environmental practices of a company or the environmental performance and positive communication about environmental performance" (TerraChoice 2010). Greenwashing refers to the promotion of seemingly ecological behaviors and attitudes and the offering of products pretending to be organic, although this is not true or contradicts the general policy of the company. In other words, greenwashing is the falsification of true information in the eyes of consumers who are deliberately misled and, for example, associate a given brand with ecological criteria, although it actually works to the detriment of the environment.

An example of building positive eco-marketing is obtaining energy guarantees of origin from individual companies. The energy guarantee of origin is part of the European Union's climate policy. It certifies, first of all, to the final consumer of this energy that it was produced from renewable energy sources or as a result of the cogeneration mechanisms used, also known as combined heat and power (CHP), which means the concurrent production of electricity or mechanical power and useful thermal energy during the same technological process. Selected examples of eco-marketing and greenwashing shows Table 21.

Table 21. Selected examples of eco-marketing and greenwashing.

Examples of eco-marketing	Examples of greenwashing
Respect for the principles of sustainable development and environmental protection – effective green marketing must concern both pro-environmental activities and care for the appropriate, environmentally friendly habits of employees – this is primarily about reducing electricity consumption, reducing the emission of harmful greenhouse gases and choosing environmentally friendly, renewable energy sources. **Eco-friendly products** – if you produce certain goods, make sure you produce them in an appropriate, environmentally friendly way, use natural materials and control both the production process and the distribution of products. **Eco-friendly packaging** – if you care about the environment and wish to implement eco marketing in your activities, also remember to choose or produce the right product packaging; in this case, biodegradable packaging or packaging made from recycled materials would be a good solution. **Educating customers** – eco-marketing is not only about saving energy and eco-friendly products or packaging – it is also a pro-social strategy that involves educating consumers about environmental protection; to increase the environmental awareness of your customers, you can organize various social campaigns, conferences or webinars or plan appropriate communication, e.g., by e-mail and regularly send an educational newsletter. **Influencer marketing** – while pursuing the eco-marketing strategy, it is also worth taking care of the promotion of pro-ecological activities – for this purpose you can, for example, establish cooperation with an influencer who also promotes ecological, environmentally friendly solutions; remember to choose the right influencer who will truly support your activities. **Social media marketing** – social media marketing is also a good way to communicate and promote green activities; plan an appropriate communication strategy on social media and show your consumers how you implement environmentally friendly activities.	**Greencrowding**, or safety in numbers. The method is based on the belief that you can hide in the crowd to avoid accusations of greenwashing. It means moving in a large group of companies and forming associations with a specific ambitious ecological goal, which is actually implemented on a small scale. **Greenlighting**, which occurs when a company emphasizes one particularly green feature of its activities or products in its communication in order to divert attention from others that are harmful to the environment. **Greenshifting** refers to companies that try to shift climate responsibility from the producer to the consumer in their communication. **Greenlabeling**, i.e., using the terminology "bio", "natural", "green" or "environmentally friendly" for non-organic products. **Greenrinsing**, which refers to a situation where a company regularly changes its ESG objectives before they are achieved. **Greenhushing**, which is the most sophisticated form of greenwashing, is a method that addresses the activities of management teams that deliberately conceal their sustainability actions and progress, or lack thereof, in order to avoid investor control.

Source: Own study.

7.4 Technological solutions related to automation, robotics and technical diagnostics in business models and strategies as an attribute of building climate responsibility

Modern engineers, creators of innovative solutions, visionaries and even futurists should think strategically but ethically. This way of thinking should be firmly rooted in the responsible behavior of society and business. Then, we become responsible, understanding at the same time the potential of this responsibility. In order to be climate-responsible at the company level, it is necessary to have the capacity to comply with the principles of efficiency and agency. This efficiency and agency should translate directly into the possibility of using different types of technological solutions treated as a core attribute of building climate responsibility. Innovative technological solutions determine the emergence of new business models. If these solutions include climate solutions, the business model used for their implementation will generate a climatic value proposition built into this business model.

Examples of technological solutions as an attribute of building climate responsibility are presented in the Table 22.

Table 22. Examples of technological solutions as an attribute of building climate responsibility.

No.	Examples of technological solutions as an attribute of building climate responsibility	Description
1.	Bioenergy	Bioenergy has complex societal and environmental interactions, including climate change feedback, biomass production and land use. The impact of bioenergy on social and environmental issues (e.g., health, poverty, biodiversity) may be positive or negative depending on local conditions and the design and implementation of specific projects. The policy context for bioenergy, and particularly biofuels, has changed rapidly and dramatically in recent years. The food-versus-fuel debate and growing concerns about other conflicts are driving a strong push for the development and implementation of sustainability criteria and frameworks. Many conflicts can be reduced, if not avoided, by encouraging synergisms in the management of natural resource, agricultural and livestock sectors as part of good governance of land use that increases rural development and contributes to poverty alleviation and a secure energy supply (Chum et al. 2011).
2.	Carbon capture and storage (carbon sequestration)	CCS stands for Carbon Capture and Storage, i.e., the technology of capturing and storing carbon dioxide. The CCS process starts by capturing CO_2 emitted by industrial processes or burning fossil fuels. Then the gas is transported (often via pipelines) to places where it is stored in specially designed geological formations, such as empty oil or gas fields. CSS has the potential to reduce CO_2 emissions from the industrial and energy sectors, but it is a technology that requires advanced infrastructure and appropriate geological conditions for the effective storage of carbon dioxide.
3.	Zero-emission energy sources Alternative fuels	Alternative fuels are fuels or energy sources that are substitutes for oil in the transport sector and contribute to its decarbonization. AFIR alternative fuels are divided into alternative fuels used in zero-emission vehicles, trains, ships or aircraft – this category includes fuels that do not cause emissions from the vehicle itself, apart from emissions resulting from fuel production (e.g., electricity).
4.	Hydrogen fuel cells and fuel cell vehicles	Hydrogen fuel cells are electrochemical devices in which hydrogen (derived from fuel) is combined with oxygen (from the air). The reaction products are electricity and heat and water (Fuel Cell Handbook 2002).
5.	Storage and energy saving	Energy storage makes it possible to prevent interruptions in the supply of electricity and heat, caused, among others, by ending fossil fuel resources and unstable energy production from renewable sources (influence of weather conditions). These technologies improve the efficiency of energy management for households, industrial application and the entire electricity system. Energy can be collected in mechanical, electrical, chemical or biological form, in the form of heat or using a magnetic field.

Source: Own study.

Exergy analysis can be used to assess the effectiveness of the technology used in terms of a system and process. Exergy analysis allows for a more comprehensive determination of the environmental impact of a given technology than the classic energy analysis and its accompanying key performance indicators (KPIs) expressed as energy efficiency or specific energy consumption per unit of product. Exergy is the minimum work required to produce the desired substance with the desired parameters from commonly occurring components of the surrounding nature, using the environment as the sole source of heat (Riekert 1974). Only a systemic approach to technological solutions aiming for positive climate change can guarantee the generation of acceptable results and their spread in the technological space of the management organization.

7.5 Digital climate technologies in business models and strategies as an attribute of building climate responsibility

Technologies, especially digital ones, change not only our reality but also habits and behaviors of a cognitive, manual and physical nature. This can have a positive effect but also a negative one, which is clearly visible, for example, in the use of mobile phones by children and adolescents. Digital climate technologies can be defined as a group of technologies that have embedded solutions based on digitalization in their structure and, on the other hand, do not generate adverse climate impact. They are usually the result of digital business models developed and implemented through a stream of long-term climate goals. Moreover, it should be noted that digital climate technologies are often a variation of previously used analog technological solutions. This opens space for reducing the operating costs of companies. In this case, a win-win strategy is the best strategy possible. The involvement of companies in the fight against climate change should be significant (Wright and Nyberg 2024). It should be based on linking the green digital transition with automation, robotics and technological diagnostics solutions.

The proposed examples of climate technologies can be used for the climate in different ways and in different places what it shows Table 23. It is important to verify their achieved climatic impact. The essence of such solutions is to cause a stream-like effect of climate efficiency.

7.6 The digital dimension of business management with reference to climate change as a source of new competitive advantage on the market resulting from technology

Digital management within current companies on the market, including the virtual market, is significantly different from analog processes. There are different business spaces, different solutions, often different products and different effects and results achieved. In order to fully master this design and implementation, it is necessary to dynamically develop digital competences and skills and develop and use the

Table 23. Examples of digital climate technologies as an attribute of building climate responsibility.

No.	Examples of digital climate technologies as an attribute of building climate responsibility	Description
1.	Advanced early warning systems for weather extremes	Machine learning technology has become a huge breakthrough in correct weather forecasting. Machine learning models far outperform traditional predictions in many ways.
2.	Spatial data to facilitate adaptation	The digitalization of spatial planning involves the processing of data on existing acts from analog to digital form and the creation of new APP data in digital form, in the course of the planning procedure. The main goal of the digitalization of spatial planning is to introduce solutions that will facilitate access to uniform, reliable and complete information on planned spatial development throughout the country.
3.	Intelligent energy management systems in urban buildings and facilities	The Building Management System (BMS) is an advanced technological solution that enables automatic control and integration of all installations in the building. Its main task is to centralize control over various systems, which allows for effective management of the facility.
4.	Intelligent mobility support systems	Connected, Cooperative and Automated Mobility (CCAM) is an essential solution for supporting smart mobility. One such solution is cooperative intelligent transport systems (C-ITS), which enable the exchange of information between vehicles and between vehicles and road infrastructure.

Source: Own study.

given potential of the virtual world for the purposes of running your own business. The COVID-19 pandemic and a number of other accompanying crises have significantly changed the business logic in many cases and significantly accelerated the process. Those organizations that overslept and did not change this logic have disappeared from the current business space. However, the point is not only that these organizations have changed this logic of doing business, but whether this logic is now a source of achieving above-average results in the sector where these organizations operate. The key question here is whether the transition to digital management for a given type of organization has led this organization to implement its processes better, faster, easier and friendlier for its customers. Have the current solutions applied by managers been appreciated by different groups of stakeholders expecting different, sometimes sophisticated, needs and values from them? In this context and interpretation, managers' understanding of the place and role of climate change and its impact on macroeconomic aspects of the functioning of the economy as well as the impact on individual sectors of the economy and individualized business models and strategies of organizations specifically functioning on the market is at the forefront. This relationship is not unambiguous and is not easy to interpret. Hence, it becomes crucial to embed climate change in systemic change management at the level of companies. In order to understand this well, managers must have a specific benefit from it. Climate change should not only be a source of market threats and the creation of new risk groups (Hewitt and Stone 2021). It should provide a platform for new dynamic strategic thinking reflected in the designed and implemented strategies of the organization (Paul et al. 2017). It will result in new products and services, which will often be climate products and services (Visscher et al. 2020). In this sense, climate change should become an accelerator and a driver of the creation of new products and services that positively affect these climate changes. This should have both global and local dimensions (Bisaro et al. 2021). It should be noted that the European Union strongly supports the development of climate services at the institutional level (Panenko et al. 2021). It is crucial that these services create multi-criteria mechanisms for monitoring climate change in different dimensions (Visman et al. 2022). The standardization of climate services and the improvement of their quality should be a priority at various decision-making levels, both public and private, non-commercial and commercial (Kolstad et al. 2021; International standard ISO 14090, 2019-06; International standard ISO 14091, 2021-02; Technical specification ISO/TS 14092, 2020-05). Therefore, it is essential to involve a wide range of decision-makers and influential organizations in shaping and developing the substantive environment of scientists and practitioners dealing with climate change (Bojovic et al. 2021). Climate change and climate services should be communicated systematically through various media. This message should be clear and reliable (Terrado et al. 2022). The result of such an approach should be an acceptable level of satisfaction of various groups of stakeholders with these products and services, which should materialize in a strong increase in the financial results of the organization. It is crucial that managers understand their impact on the natural ecosystem, which depends on the economic and social ecosystem created by local communities and various types of economic operators. In thought processes, there needs to be a strong correlation between science and practice and vice versa (Coen 2021; Bray and Storch 2009). To harness the potential of climate change, today's managers need to reshape their long-term and short-term thinking and transform their organizations based on the assumption that climate change is a huge opportunity rather than a threat. In order to be able to do this, it is necessary to accurately identify the phenomena in the social and market environment based on the principles of early warning systems. This applies to the strategic, tactical and operational levels of the organization. Climate change should become a factor that even changes the organizational structures of various types of organizations (Baccarani et al. 2021, pp. 640–661).

7.7 The impact of climate change on the strategic and tactical–operational functioning of companies and business digitalization

The climate impact at the company level is already very noticeable. It may not yet be the effect that many supporters of grassroots work for the climate and its dynamic changes would like, but there is great progress and light at the end of the tunnel towards the increased dynamics of economic and social changes within climate adaptation strategies. The proper and precise understanding by managers of the impact of climate change on the strategic functioning of companies can only be expressed by changing the structure and qualitative nature of the long-term goals of the organizations built in its strategy. It may also involve changing the logic and methodology of revenue generation and offering a new environmentally friendly value proposition built into the business model for climate-conscious customers of a given organization. This value proposition in the business model can be expressed in the creation of unique, rare and innovative climate services (Larosa and Mysiak 2020). It should be noted that at present, in the situation of the existence of companies in the environment of numerous crises, defining the long-term goals of the organization and striving for their maturity is very difficult, and in some cases even impossible. Hence, the assumptions of the strategy concept combining aspects of strategy and tactics at the same time with a focus on emerging business opportunities are used. In view of the above, the creation and implementation of long-term goals of the organization are now a set of micro-goals based on the emergence of further micro-decisions supported by a set of micro-corrections. If in every micro-problem solved, in every micro-decision, every micro-goal and every micro-correction there is a climatic factor embedded and the way of its solution, then automatically the result of the entire strategy of the organization will be a functional strategy of creating a positive climate impact. This will be done while generating financial value for an organization aware of the climate problems that affect the world. It will also be consistent with the expectations of all groups of communities, especially local communities. It is also crucial here that companies build trust towards local communities sensitive to climate change (Lacey et al. 2018). Then, the principles of climatic social impact are being shaped, the recipients of which are broad groups of local communities. A proper understanding of climate change at the tactical level involves the reconfiguration and reconstruction of its processes and projects, where these two ontological entities have a different result than before. They result in climate products and services, or products that have a positive impact on the climate, as well as services that shape a positive impact and result for the climate that builds trust in the organization. In this case, trust management in the organization focused on trustworthy products and services is very important. In order to characterize processes and projects with features and properties that create climatic values, they should be qualified and categorized. This can be achieved by making primary assumptions relevant to the achieved results embedded in the products of a given organization consistent with the logic, methodology and conceptual operation suitable for achieving a positive climate impact. The digitalization of the entire value chain can help to effectively and efficiently manage processes with the goal of achieving climate efficiency. For example, the use of a digital twin of the entire value chain can allow for its thorough and flexible redevelopment in areas such as logistics, production cost management, price management and production design. Appropriate management of the sustainable digital footprint, considered the sum of the positive and negative social, environmental and economic impacts associated with the use of new technologies in the organization's activities along the entire value chain, becomes important here. The goal of sustainable digital footprint management is to create the right value for stakeholders and contribute it to the sustainable development of the organization, thanks to the effective use of new digital technologies in current business activities.

A proper understanding of climate change at the operational level involves adopting the assumptions of a circular economy, i.e., one in which no waste is generated, or if it is generated, it becomes semi-finished products for other technological processes. In the operational area, the climatic criteria should be embedded in the applied technology of products and services and in the specific selection of raw materials and semi-finished products with properties for re-use or return to the modelled technological process. Waste-free management then becomes a mechanism, not only for cost optimization, but above all a platform for generating additional revenues. The core components that contribute to the implementation of the zero-waste philosophy are reliable data on waste, awareness and education, behavioral change, public participation, optimal design of recycling and recovery processes, product co-responsibility, rules and regulations, producer and consumer responsibility, cradle-to-cradle projects and creating a market for waste (Zaman 2014, p. 683). Then, a systemic cost optimization occurs at the operational level in compliance with the ecological criteria adopted for application. Low-emission or zero-emission operational processes are a determinant of environmental and climate efficiency.

7.8 Digital business innovation as a source of creating products and services rich in climate values

The digitalization and virtualization of the space–time of business and the economy opens new, innovatively oriented layers of creative thinking in the technological and operational context. Technological imagination focused on digital innovation generates unprecedented solutions that often amaze even the biggest opponents of the progress of digitalization. In order to be able to produce a product or service rich in climatic values, it becomes crucial to create climate innovations at the level of the product or service itself and at the levels of technology, organization and management. The mechanisms adopted to develop digital technological innovations can be used for this purpose. While digital innovation is a means for new entrants to use digital technology to challenge companies already on the market, which ultimately results in a radical transformation at the industry level, it also provides incumbents with opportunities to expand and extend their product and service portfolio to new domains. However, a main challenge for any company trying to manage digital innovation is to understand the unique properties of digital innovation processes (Yoo et al. 2010). The effects of digital innovation are particularly evident for companies that deal with information-based products that can be fully digitalized (Nylén and Holmstrom 2015). This is the case in climate change processes. In this thought process, innovation, understood as a change for the better, shapes the end result in the creation of a whole network of new products and services that generate new positive climatic values in their genotype itself. Well-constructed assumptions and boundary conditions generate a product and/or service that is suitable for expectations and rich in climatic values. It should be noted that climate-related services and products are particularly important in the context of a new climate-sensitive business adaptation

(Dolsak et al. 2018; Gancheva et al. 2022). It is this area of entrepreneurial activity that should be particularly focused on activities aimed at creating new climate innovations. Climate innovation can be a powerful weapon in creating a positive dimension of climate change implemented by entrepreneurs. Climate innovation in climate services can be created by means of artificial intelligence. In this case, the so-called climate algocracy, generating new climate services, can be used. In the field of new climate products, technologies resulting from the principles of Industry 4.0 and 5.0 may be used, supported by solutions resulting from the area of automation and robotics. In the area of the circular economy, the principles of the Internet of Things can be used, as can the logic of monetization of business and micropayments in the processes of digitization. This can be supported by the principles of the sharing economy in order to reduce the use of resources by taking advantage of their redundancy. These selected, aggregated areas of innovation can be a powerful generator of the polarization of the economy from its adverse impact on the climate towards the creation of positive climate change embedded precisely in these proposed solutions. The combined, twin digital and climate transformation is creating a new technological space for society. Socio-technical approaches to transformation research focus on the role of technological innovation in triggering transformation, but emphasize that technology can only provide a function to society if embedded in a socio-technical system. This system is created by a network of actors such as users, producers, suppliers, public authorities and financiers, as well as the relevant infrastructures, behavioral patterns, cultural values and policies that link them (Elzen 2004). An example of digital business innovation as a source of smart climate value-rich products and services is the combination of satellite-based monitoring and artificial intelligence, which is revolutionizing the way we monitor climate change. Satellites provide global communication and precise GPS navigation. They enable accurate weather forecasts and Internet access in remote areas of the world. The future use of satellite technologies includes, inter alia, the development of a constellation of low-orbit satellites and the integration of artificial intelligence in this area. Telecommunication, navigation and meteorological satellites are widely used in climatology. Combining drones as cyber-physical systems with artificial intelligence has great potential for the systemic monitoring of climate change. In agriculture, for example, drones can be used to monitor crops, identify core areas in need of irrigation or protect against pests. Artificial intelligence and smart geolocation solutions can also be effectively used to measure and monitor climate anomalies and extremes, providing multidimensional strategic information on critical climate points and creating mechanisms for the appropriate management of these points. Geospatial artificial intelligence (GeoAI) is the application of artificial intelligence (AI) combined with geospatial data, science and technology to dynamically accelerate decision-making processes based on real technological capabilities. Climate innovation can also be linked to or directly driven by cloud-based technologies. Cloud services also include Platform as a Service (PaaS), for example, application structure, and Infrastructure as a Service (IaaS), which provides technical infrastructure components such as storage, the processor, memory and network (Chard et al. 2010). Cloud computing services should be strongly developed both in terms of technology and purposeful content giving final value for positive climate change.

7.9 Digital business models and shaping climate responsibility

Is there a revolution or an evolution in the digital world of economy and business? This is a very important question which is difficult to answer unambiguously and clearly. Revolution creates breakthroughs and is a drastic catalyst for dynamic changes, and is sometimes difficult to control. Evolution gives breathing space to evaluate, verify and validate this process. The revolutionary digital business that arises from the impact of these technologies on markets explores the concepts and skills that managers need to manage this type of business (Sousa and Rocha 2019, pp. 257–263). Digital business operates in a digital ecosystem. Digital ecosystems are driven by new digital technologies that have transformed the very nature and extent of traditional interdependencies. Digital ecosystems are much more extensive and their reach and importance go beyond traditional value chains and conventional industrial structures (Subramaniam et al. 2018). The following approach can be taken into account when designing digital business models:

1. The use of digital business model patterns as a first access point for devising the initial ideas and a selection of certain patterns for further analysis.
2. Reflection on strategic intent, creating clusters from the idea design process while streamlining this with an eye for possible synergies.
3. The use of value drivers to clarify how technology is employed in the implementation of digital business model ideas (Blaschke et al. 2017). Digital business models or processes save resources. The use of Internet-based technologies increases the number of operations and procedures, leading to flexible production, automatic administration or even a shortened marketing phase. Saving costs, time and resources improves the cost structure of the company in general (BarNir et al. 2003). It is interesting to note that companies will not have to relocate their production to developing countries to exploit their cheap labor resources, as they can achieve the same or lower costs by using AI technologies (Makridakis 2017).

Digital platforms are a response to strategic changes in the business area caused by the dynamic digital transformation. There are one-sided, two-sided and multi-sided platforms that can apply different types of business monetization. It all depends on the willingness to establish the appropriate relationship with customers. Some examples of monetization methods include the freemium model, in-app ads, in-app purchases, and subscription models. The key is that the logic of digital business is completely different from analog business. The climate solutions resulting from the business models used may therefore have analog, digital and hybrid dimensions, which are a compilation of analog-to-digital and digital-to-analog solutions. In the case of analog solutions, the aspect of constructing linear business models is primarily used, and in the case of digital solutions, the circular aspect of business models is often used. The dimension of the value chain for analog and digital solutions is different. The situation is similar in economic and financial terms, where in analog solutions there is a classic financial account and, for example, in relation to the monetization of digital business models, e.g., digital platforms, it is the calculation of marginal costs. The term 'the economics of digital platforms' are used in this case. The optimal solution for shaping climate responsibility is to choose the right types of digital platforms for the right types of business monetization in order to achieve the accepted dynamics of achieving organizational results. With regard to the assumptions of climate responsibility, digital platforms can be treated as a specific climate ecosystem for the common market, where climate criteria are noticeable and recognized. The concept of the platformization of the principles of climate responsibility can be strongly applicable here, where the following hybrid solutions can be used:

- from physical linear to digital linear,
- from digital linear to digital multi-sided,
- from physical linear to digital multi-sided,
- from linear multi-sided to digital multi-sided.

This process clearly shows that the appropriate configuration of the digital business model, the digital platform in combination with the adopted appropriate digital and hybrid solutions and in combination with the right types of monetization can produce surprising climate results at the level of climate-responsible companies.

7.10 Responsible Artificial Intelligence in the processes of building climate responsibility

Can Artificial Intelligence, as a technological intelligence, be responsible? If so, who can be and is responsible for this, and therefore can also be held responsible if the results are negative, for example, for humanity?

Artificial Intelligence as technology has numerous applications in the industry (Brynjolfsson and McAfee 2014). The term Artificial Intelligence (AI) refers to computers that mimic aspects of human thinking. A simple electronic calculator does not have AI status, but a machine that can learn from its mistakes, or that can exhibit reasoning, has the status of AI. There is no exact dividing line between these extremes (Gibilis 1994). Artificial Intelligence is a field of science that deals with solving problems that are effectively non-algorithmizable, based on knowledge models (Spirit 1997). Artificial Intelligence solutions are revolutionizing the broad picture of social and economic reality. They provide a dynamic platform for an innovative stream of emerging technological, organizational and management capabilities in various types of organizations, while accelerating the achievement of appropriate results in various fields of business existence and operation of various types of processes and projects. This also includes aspects of climate change and related climate responsibility. Artificial Intelligence and the development thereof should be responsible. In 2021, on the basis of the resolution of the 41st session of the General Conference (41 C/Resolution), UNESCO adopted a Recommendation on the Ethics of Artificial Intelligence, the first and so far the only legal instrument of global scope dedicated to this issue. Its application enhances opportunities for both the development of national policies and international cooperation and dialogue. On 19 February 2020, the European Commission published its White Paper on Artificial Intelligence: a European approach to excellence and trust and presented a report on the safety and liability implications of artificial intelligence, the Internet of Things and robotics to the European Parliament. The Commission's White Paper on Artificial Intelligence sets out a clear vision of artificial intelligence in Europe as an ecosystem of excellence and trust (European Commission document COM 2020). The responsibility of Artificial Intelligence has technological and informational dimensions. On the one hand, it is about the so-called morality of algorithms and ethics of machines, robots and cyborgs, while on the other hand, it is about, for example, the quality of the provided informative climate services related to climate forecasting and dysfunctions, anomalies and climate extremes. So, in this approach, technology determines the quality of information and information, and initiates technical and organizational–management processes with the goal of achieving appropriate effectiveness and climate efficiency. Responsible AI Principles are referred to in the EU documentation as a strategic document: Artificial Intelligence for Europe (Communication from the Commission to the European Parliament, the European Council, the European Economic and Social Committee and the Committee of the Regions 2018). What is important here is the entire, comprehensive ecosystem of Artificial Intelligence. The AI ecosystem is a dynamic and spatial environment open to and focused on initiating and supporting activities undertaken by a wide range of actors of the digital network for the development of innovation in the field of AI. The goal of its development is the strong positioning of intellectual property in the global value chain related to data processing. Its dynamics and design serve to minimize the possible risks associated with AI creation and the ongoing transformation of the economy and society. Climate data must be managed responsibly. Data must be collected, shared and managed in a way that maintains trust and takes due account of privacy and security. Some data that are relevant for climate applications can be very sensitive in terms of privacy. There is a risk that if AI projects in the climate space are irresponsible when it comes to accessing or processing data, they could undermine trust in AI in climate-related sectors (Climate Change and AI Recommendations for Government Action 2021). Therefore, responsibility and ethics in the application of various solutions of Artificial Intelligence should resound very strongly at the center of the scientific discussion, because only then will we be able to positively change our world, including the climate, taking into account modern technologies.

7.11 Climate products and services supported by digital solutions as a result of the multidimensional operation of companies and other organizations

A product, a service, a business model, and a strategy are the core components of business for the existence of any company. Companies operating on the market today compete not only by means of products and services, but above all with business models. It is the correct configuration of the business model that generates the expected financial result of the organization. This configuration is the sum of coherent components that generate the expected value. This value for today's economic conditions may include a specific triangulation of relationships aimed at a synergistic, symmetrical and symbiotic dependence system generating social and climatic economic value. The economic dimension should depend on the extent of adaptation in the area of climate change (International standard ISO 14093). Such a result should be built into climate products and services. It is important to be aware that currently climate products and services are mainly interpreted as information services resulting from forecasting and anticipating potential climate change (Jebeile and Roussos 2023). This also includes a comprehensive collection of data and knowledge on the state of the climate worldwide and making multidimensional comparisons of analyses (Boon et al. 2022). They are used for diagnosis, synthesis and inference aimed at anticipating and combating climate change (Bessembinder et al. 2019). In the presented approach, this range of climate products and services is definitely widened. In this interpretation, the climate product and service also includes all products and services that noticeably change a given sector of the economy. They do this by eliminating or substituting other products and services that have a negative impact on the environment and the climate or are only climate-neutral at best. In cognitive terms, the idea embedded in the content concerns the substitutability of products and services adversely affecting the climate with products and services positively affecting it in a dynamic way, reversing negative trends in the economy, society and business, unfortunately often aimed at the conscious generation of climate hazards and catastrophes. Then, organizations creating the socio-ecological ecosystem (Hewitt et al. 2020) become climate companies. A climate company can therefore be defined as a company that undertakes decision-making processes at the strategic and tactical–operational level with a focus on climate values shared by all members of the organization, and its development processes are conducted in accordance with the principles of sustainable business and development. Then, the process of climate value monetization can take place through the climate company. This enables a mutual economic and climate benefit in the climate company.

Core climate products and services supported by community-based climate responsibility include:

- in the area of environmental engineering and climate engineering – environmental applications and assessments, environmental impact assessments, sectoral permits for the use of the environment, integrated permits, inventories and environmental expertise, environmental quality studies and the design and co-financing of investments,
- in the area of climate change – climate change risk assessments, plans and projects for adaptation to climate change, simulations of climate impacts precisely for the analyzed objects and areas, concepts of effective adaptation and mitigation solutions based on nature, construction and technical projects with construction agreements, cost estimates and technical specifications for the execution of works, execution of construction and implementation works, implementation of blue and green infrastructure (bioretention technologies) and care for and supervision over the adaptation of green infrastructure.

7.12 Gamification in the processes of shaping climate responsibility

Simulation games that describe and utilize climate change in their operational scenarios can be an effective and efficient tool for multidimensional climate education and thus shape positive attitudes of climate responsibility, especially among young people. Inspiring young people to undertake positive climate and environmental behavior should be a priority at the level of educational solutions that shape young people's attitudes both formally and informally. In this way, interactive processes of teaching climate values to young people can be applied (Makrakis et al. 2013, pp. 54–72). This is all the more important as young people consider the climate and its changes from the perspective of their and their children's long-term existence in a very dynamically changing world. Their life perspective is longer and they are the biggest advocates of shaping smart climate values in various fields of their professional and private lives. A strong commitment to combining work and leisure in a sustainable way is a strong catalyst for and accelerator towards positive climate change. This potential should be recognized, appreciated and used for a new dimension of climate adaptation for businesses, regions and cities. In addition, elements of gamification using computer games can also be applied in this process, triggering the imagination of young people to build a climate-positive world in a virtual space, which can then be transferred to the real world. Gamification solutions should always be based on modern climate competences.

7.13 Conclusions

Proper interpretation and application of the principles of climate responsibility in the operational and technological management of companies in relation to their business models allows for the proper use of the potential inherent in the principles of digital transformation, automation, robotics, diagnostics and artificial intelligence. Such a wide range of modern and innovative solutions becomes an incredible platform for creative thinking by many managers and entrepreneurs. Then, technological management creates an intellectual space for opening strategic paths for developing climate responsibility mechanisms with the use of technology and business digitalization at the company level.

References

41 C/Resolution 34: https://unesdoc.unesco.org/ark:/48223/pf0000380399.

Baccarani, C., Brunetti, F. anf Martin, J. (2021). *Climate change and organizational management: toward a new paradigm*. The TQM Journal Vol. 33 No. 3, pp. 640-661 © Emerald Publishing Limited 1754-2731, DOI 10.1108/TQM-05-2020-0094.

BarNir, A., Gallaugher, J.M. and Augner, P. (2003). *Business process digitization, strategy, and the impact of firm age and size: the case of the magazine publishing industry*, Journal of Business Venturing Vol. 18, issue 6: 789ff.

Bessembinder, J., Terrado, M., Hewitt, Ch., Garrett, N., Kotova, L., Buonocore, M. and Groenland, R. (2019). *Need for a common typology of climate services,* Climate Services, Volume 16, 100135, https://doi.org/10.1016/j.cliser.2019.100135.

Bisaro, A., Hinkel, J., Le Cozannet, G., van der Pol, T. and Haas, A. (2021). *Global Climate Services: A Typology of Global Decisions Influenced by Climate Risk.* Front. Mar. Sci. 8: 728687. doi: 10.3389/fmars.2021.728687.

Blaschke, M., Cigaina, M, Riss, U.V. and Shoshan, I. (2017). *Designing Business Models for the Digital Economy*, Springer International Publishing Switzerland 2017, Oswald, G. and Kleinemeier, M. (eds.). Shaping the Digital Enterprise, DOI 10.1007/978-3-319-40967-2_6.

Bojovic, D., St. Clair, A.L., Christel, I., Terrado, M., Stanzel, P., Gonzalez, P. and Palin, E.J. (2021). *Engagement, involvement and empowerment: Three realms of a coproduction framework for climate services,* Global Environmental Change, Volume 68, 102271, https://doi.org/10.1016/j.gloenvcha.2021.102271.

Boon, E., Wright, S.J., Biesbroek, R., Goosen, H. and Ludwig, F. (2022). *Successful climate services for adaptation: What we know, don't know and need to know*, Climate Services Volume 27, 100314, https://doi.org/10.1016/j.cliser.2022.100314.

Bray, D. and von Storch, H. (2009). *"Prediction" or "Projection"?: The Nomenclature of Climate Science.* Science Communication, 30(4): 534–543. https://doi.org/10.1177/1075547009333698.

Brynjolfsson, E. and McAfee, A. (2014). T*he second machine age: Work progress, and prosperity in a time of brilliant technologies*. New York: W.W. Norton & Company.

Chard, K., Caton, S., Rana, O. and Bubendorfer, K. (2010). Social cloud: cloud computing in social networks. Proceedings of the 3rd international conference on cloud computing IEEE cloud.

Chum, H., Faaij, A., Moreira, J., Berndes, G., Dhamija, P., Dong, H., Gabrielle, B., Goss Eng, A., Lucht, W., Mapako, M., Masera Cerutti, O., McIntyre, T., Minowa, T. and Pingoud, K. (2011). *Bioenergy. In IPCC Special Report on Renewable Energy Sources and Climate Change Mitigation*, Edenhofer, O., Pichs-Madruga, R., Sokona, Y., Seyboth, K., Matschoss, P., Kadner, S., Zwickel, T., Eickemeier, P., Hansen, G., Schlömer, S. and von Stechow, C. (eds.), Cambridge University Press, Cambridge, United Kingdom and New York, NY, USA.

Climate Change And Ai Recommendations for Government Action. (2021). Global Partnership on AI Report. In collaboration with Climate Change AI and the Centre for AI & Climate, Issued: November 2021.

Coen, D.R. (2021). *A brief history of usable climate science.* Climatic Change 167, 51 https://doi.org/10.1007/s10584-021-03181-2.

Communication from the Commission to the European Parliament, the European Council, the European Economic and Social Committee and the Committee of the Regions. (2018). Artificial Intelligence for Europe, {SWD137 final}

Delmas, M. and Burbano, V. (2011). *The drivers of greenwashing.* Calif Manag Rev 54(1), pp. 64–87. https://doi.org/10.1525/cmr.2011.54.1.64.

Digital Deloitte. (2017). Adopting the power of conversational UX Chatbots, https://www2. deloitte.com/pl/pl.html.

Dolsak, N. and Prakash, A. (2018). *The Politics of Climate Change Adaptation*, Annual Review of Environment and Resources, 43: 317–41, https://doi.org/10.1146/annurev-environ-102017-025739.

Duch, W. (1997). *Fascynujący świat komputerów* [The fascinating world of computers], Wyd. Nakom.

Elzen, B., Geels, F.W. and Green, K. (2004). *System Innovation and the Transition to Sustainability: Theory, Evidence and Policy.* Edward Elgar Publishing, Cheltenham, UK; Northhampton, MA, USA.

European Commission Document COM. (2020). White Paper on Artificial Intelligence – A European approach to excellence and trust. 65, final, Brussels 109.02.2020.

Fuel Cell Handbook. (2002). Sixth edition, EG&G Technical Services, Inc. Science Applications International Corporation, DOE/NETL- 2002/1179.

Gancheva, M., Lundberg, P. and Vroom, I. (2022). *Climate adaptation: Measuring performance, defining targets and ensuring sustainability*, European Committee of the Regions.

Gibilis, S. (ed.), (1994). The McGraw-Hill Illustrated Encyclopedia of Robotics & Artificial Intelligence.

Härting, R.Ch., Reichstein, Ch. and Schad, M. (2018). *Potentials of Digital Business Models – Empirical investigation of data driven impacts in industry*, Procedia Computer Science 126: 1495–1506, (https://creativecommons.org/licenses/by-nc-nd/4.0/.

Hewitt, C. D., Guglielmo, F., Joussaume, S., Bessembinder, J., Christel, I., Doblas-Reyes, F. J., Djurdjevic, V., Garrett, N., Kjellström, E., Krzic, A., Máñez Costa, M. and St. Clair, A. L. (2020).

Recommendations for Future Research Priorities for Climate Modeling and Climate Services, ©2021 American Meteorological Society, https://doi.org/10.1175/BAMS-D-20-0103.1

Hewitt, C.D. and Stone, R. (2021). *Climate services for managing societal risks and opportunities*, Climate Services, 23: 100240, https://doi.org/10.1016/j.cliser.2021.100240.

Hewitt, C., Mason, S. and Walland, D. (2012). *The Global Framework for Climate Services*. Nat. Climate Change, 2: 831–832, https://doi.org/10.1038/nclimate1745.

International standard ISO 14090 First edition 2019-06, *Adaptation to climate change — Principles, requirements and guidelines.*

International standard ISO 14091 First edition, 2021-02, *Adaptation to climate change — Guidelines on vulnerability, impacts and risk assessment.*

International standard ISO 14093 First edition 2022-1, *Mechanism for financing local adaptation to climate change — Performance-based climate resilience grants — Requirements and guidelines.*

Jebeile, J. and Roussos, J. (2023). *Usability of climate information: Toward a new scientific framework*, WIREs Clim Change.14: e833., https://doi.org/10.1002/wcc.833.

Kolstad, E.W., Sofienlund, O.N., Kvamsås, H., Stiller-Reeve, M.A., Neby, S., Paasche, Ø., Pontoppidan, M., Sobolowski, S.P., Haarstad, H., Oseland, S.E., Omdahl, L. and Waage, S. (2019). *Trials, errors, and improvements in coproduction of climate services*, In final form 1 April 2019©2019 American Meteorological Society, DOI:10.1175/BAMS-D-18-0201.1.

König, M. (2016). *Different patterns in the evolution of digital and non-digital ventures' business models*, Technological Forecasting & Social Change (2018), https://doi.org/10.1016/j.techfore.2018.05.006.

Lacey, J., Howden, M., Cvitanovic, C. and Clovin, R.M. (2018). *Understanding and managing trust at the climate science–policy interface*. Nature Clim Change 8: 22. https://doi.org/10.1038/s41558-017-0010-z.

Lam, J.S.L. and Li, K.X. (2019). *Green port marketing for sustainable growth and development. Transport Policy*, 84: 73–81. doi:10.1016/j.tranpol.2019.04.011.

Larosa, F. and Mysiak, J. (2020). *Business models for climate services: An analysis*, Climate Services 17, 100111, https://doi.org/10.1016/j.cliser.2019.100111.

Makrakis, V., Larios, N. and Kaliantzi, G. (2013). ICT-Enabled Climate Change Education for Sustainable Development Across the School Curriculum, Journal of Teacher Education for Sustainability, 14(2): 54–72.

Makridakis, S. (2017). *The forthcoming Artificial Intelligence (AI) revolution: Its impact on society andfirms*, FuturesVol. 90: 46–60.

Nylén, D. and Holmstrom, J. (2015). *Digital innovation strategy: A framework for diagnosing and improving digital product and service innovation*, Business Horizons 58: 57—67, http://dx.doi.org/10.1016/j.bushor.2014.09.001.

Orlove, B. (2022). *The Concept of Adaptation*, Annual Review of Environment and Resources, 47: 535–81, https://doi.org/10.1146/annurev-environ-112320-095719.

Panenko, A., George, E. and Lutoff, C. (2021). *Towards the development of climate adaptation knowledge-action systems in the European Union: An institutional approach to climate service analysis*, Climate Services Volume 24, 100265, https://doi.org/10.1016/j.cliser.2021.100265.

Papadas, K.K., Avlonitis, G.J. and Carrigan, M. (2017). Green marketing orientation: Conceptualization, scale development and validation. Journal of Business Research, 80: 236–246. doi:10.1016/j.jbusres.2017.05.024.

Paul, A., Lang, J.W.B. and Baumgartner, R. J. (2017). *A multilevel approach for assessing business strategies on climate change*, Journal of Cleaner Production 160: 50–70, http://dx.doi.org/10.1016/j.jclepro.2017.04.030.

Riekert, L. (1974). *The Efficiency of Energy Utilization in Chemical Processes*. Chem. Eng.Sci. 1974, No 29.

Sousa, M.J. and Rocha, A. (2019). *Skills for disruptive digital business*, Journal of Business Research 94 (2019) pp. 257–263, https://doi.org/10.1016/j.jbusres.2017.12.051.

Subramaniam, M., Iyer, B. and Venkatraman, V. (2018). *Competing in digital ecosystems*, Business Horizons https://doi.org/10.1016/j.bushor.2018.08.013.

Technical specification ISO/TS 14092 Reference number ISO/TS 14092:2020(E) First edition 2020-05, *Adaptation to climate change — Requirements and guidance on adaptation planning for local governments and communities.*

TerraChoice. (2010). *The sins of greenwashing: home and family edition.* http://sinsofgreenwashing.org/findings/the-seven-sins/.

Terrado, M., Calvo, L. and Christel, I. (2022). *Towards more effective visualisations in climate services: good practices and recommendations.* Climatic Change 172, 18, https://doi.org/10.1007/s10584-022-03365-4.

Visman, E., Vincent, K., Steynor, A., Karani, I. and Mwangi, E. (2022). *Defining metrics for monitoring and evaluating the impact of co-production in climate services,* Climate Services Volume 26, 100297, https://doi.org/10.1016/j.cliser.2022.100297.

Visscher, K., Stegmaier, P., Damm, A., Hamaker-Taylor, R., Harjanne, A. and Giordano, R. (2020), *Matching supply and demand: A typology of climate services*, Climate Services, Volume 17, 100136, https://doi.org/10.1016/j.cliser.2019.100136.

Walker, K. and Wan, F. (2011) The harm of symbolic actions and green-washing: corporate actions and communications on environmental performance and their financial implications. J. Bus. Ethics 109(2): 227–242. https://doi.org/10.1007/s10551-011-1122-4.

Wright Ch. and Nyberg, D. (2024). *Corporations and climate change: An overview*, WIREs ClimChange. 2024; 15: e919. wires.wiley.com/climatechange, https://doi.org/10.1002/wcc.919.

Xu, Y. (2018). Digital business model effectuation: An agile approach, Computers in Human Behavior, https://doi.org/10.1016/j.chb.2018.10.021.

Yoo, Y., Lyytinen, K.J., Boland, R.J.Jr. and Berente, N. (2010). *The next wave of digital innovation: Opportunities and challenges: A report on the research workshop 'digital challenges in innovation research.'* Retrieved from http://papers.ssrn.com/sol3/papers.cfm?abstract_id=1622170.

Zaman, A.U. (2014). *Identification of key assessment indicators of the zero waste management systems, Ecological Indicators*, p. 683. Zaman, A.U. (2014), Measuring waste management performance using the 'Zero Waste Index': the case of Adelaide, Australia, Journal of Cleaner Production, Zaman, A.U. (2016), A comprehensive study of the environmental and economic benefits of resource recovery from global waste management systems, Journal of Cleaner Production 124: 41–50. Zaman, A.U. (2022), Zero-Waste: A New Sustainability Paradigm for Addressing the Global Waste Problem, in: The Vision Zero Handbook: Theory, Technology and Management for a Zero Casualty Policy.

8

The Author's Original Methodology and the Concept of Building Climate-Changing Business Models

8.1 Introduction

The aim of this chapter is to present the author's original methodology and concept of building climate-changing business models. The scope of the chapter includes the adopted research methodology for building the configuration of business models creating climate services and the principles of building components of configurations of business models that positively change the climate. The subject of the chapter is to clarify and create a proper understanding of the potential of building climate-changing business models, thus eliminating the negative impact that numerous companies have on the climate and the environment.

8.2 The adopted research methodology for building the configuration of business models creating climate services

A business model is a representation of the actual state of the market and its customers. To achieve this, the designed mapping must work in the practice of business decision-making. The path from building a business model to achieving the predicted results is very difficult, complex and often the result of a large number of attempts, mistakes and successful or unsuccessful experiments. As regards the emergence of new climate services, attention should be paid to their compatibility with the business models used. It is becoming increasingly important to link the new climate services offered with the implemented business models of various

DOI: 10.1201/9781003474401-8

types of organizations, including companies. This can be accomplished through new proposals for climate values for different stakeholder groups and the income generation logic required for these organizations. In order to define the principles of building the configuration of business models creating climate services, it is crucial to determine the logical and justifiable optimal methodology. A five-step conceptual process of building the configuration of business models creating climate services was used here and is presented in the Table 24.

Table 24. A five-step conceptual process of building the configurations of business models creating climate services.

No.	Research Activity
1.	Determining the taxonomy and typology of climate services with a brief description
2.	Indicating selected climate services positively impacting climate change
3.	Determining the taxonomy and typology of business models and components of business models creating climate services together with their short description
4.	Proposing a universal canvas for business models creating climate services
5.	Linking the type of climate service with core business models and components of business models creating climate services for achieving optimal values and benefits generated by climate services for society, the economy and business

Source: Jabłoński A. (2023, p. 144).

The proposed five-step, consecutive conceptual process of building the configuration of business models creating climate services is vitally important in the current socio-economic ecosystem characterized by various types of crises which strongly affect the existence of various types of organizations. Often, organizations have significant problems with the consistency between the services offered and the business models allocated to the market in the form of an appropriate climate value proposition.

The lack of this complete and clear consistency is reflected in the loss of a number of market opportunities and, consequently, many limitations in positive cash flows. Such a competency gap in the management of organizations should be eliminated by managers in a systemic way, based on the continuous verification of the changing needs and preferences of customers and adjusting the designed and subsequently operated business models to them.

8.3 The component-based construction of configurations of business models that positively change the climate

The integrity, tightness and component consistency of a business model ensure its functionality and thus its feasibility and efficiency. The structure of business model components should be highly consistent in the context of the purpose function specified for it. The dependent arrangement of component relationships should include both components aimed at the interior of the organization and open to the external business ecosystem. A business model describes a firm's logic of creating, delivering, and capturing value (Teece 2010, pp. 172–194). Business models are

Table 25. Selected types of business models with a positive impact on climate change and selected definitions.

No.	Type of business models with a positive impact on climate change	Selected definition of the business model
1.	Sustainable Business Model (Jabłoński 2013a)	A sustainable business model that builds the long-term value of a socially responsible company is a model built through the combined use of the concepts of corporate social responsibility and value-based management, which ensures that the needs of shareholders and other stakeholder groups are met, through the skillful balancing of the company's potentials, towards the generation of value *distributed in a sustainable way, enabling the continuity of business management. According to this* concept, the sustainable business model is a hybrid model, i.e., a model built in a subject–object system. The components of this model are entities centered around the company creating mutual relationships, influencing company value drivers and strategic factors related to the theory of corporate social responsibility, value-based management, stakeholder theory, and shareholder theory that function in a mutual relationship based on the principles of sustainability. This model is a holistic model with a reduced character, which can be used in various sectors of the economy treated as a subsystem of the entire ecosystem. This means that this model and its construction are located in the medium-range theory.
2.	Circular Business Model	A circular business model is the rationale of how an organization creates, delivers and captures value with and within a closed material loop (Mentink 2014).
3.	Social Business Model	The specificity of social business models is based on favoring social profit-oriented shareholders and clearly specifying the social profit objective (Yunus et al. 2010).
4.	Progressive Business Model	A progressive business model (PBM) is not a conscious project. Instead, it is the result of an evolutionary interference of regulation and legislation that severely limits the independence of business. PBM uses legislation/regulation to instruct and control major business decisions to implement a set of climate improvement targets (Rochlin 2021).
5.	Inclusive Business Model	An inclusive business model is a way of creating, delivering and capturing value, providing access to resources (e.g., finance, technology, market channels) – as well as creating decision-making space for their use (Rosenstock et al. 2020).
6.	Adaptive Business Model	An adaptive business model is a model that incorporates the structure and processes of a mutual partnership underpinning value creation so that it effectively delivers and captures change as it responds to it through internal or external shocks (Rosenstock et al. 2020).

7.	Green Business Model (Jabłoński 2013b)	The "green" dimension of a company's operation, which includes the organization as a whole, builds its reputation. A company managed in this way takes into account the expectations of the "silent stakeholder", which is the environment represented by customers with ecological values, the community at global, national and local levels, governmental and non-governmental organizations setting both legal and social standards that take into account ecological criteria, and any other entity for which the ecological factor is significant. The company constructs a business model and strategy that will guarantee the projected business results by shaping the competitive potential guaranteeing continuous achievement of an unsustainable competitive advantage, which can be understood as obtaining significantly higher financial results in relation to competing organizations offering similar value to customers. Considering the dimension of the functioning of the company in the context of the business model and strategy, we are dealing with many such dimensions. The selected dimensions include: (1) The green strategic dimension of business eco-management; (2) The green process dimension through the company's green value chain; (3) The green resource dimension of the company's operation based on green resources; (4) The green dimension of the development of the company's intellectual capital based on the creation of ecological knowledge and competences; (5) The green product dimension of the company's operations based on eco-innovative products; (6) The green subjective dimension based on the fulfilled needs of the company's green stakeholders.
8.	Technological Business Model	A technological business model can be presented as a business model of which the core distinguishing components are components that are an innovative technology, its part and/or its configuration, generating higher value than other configurable components. The configuration of the technological business model provides unique value to customers in the form of a product and/or service offered through modern mechanisms of offering this value. This business model is able to increase revenue exponentially. The most common environment for technological business models is the network environment and virtual space. Analyzing the stages of designing and operationalizing the technological layout of the business model in detail, the logic indicating the presence of milestones that bring the company closer to achieving the assumed performance is important. In terms of design, this is particularly important for defining the core attributes of the company's technological business model, with particular emphasis on attributes resulting from technology. While the operationalization of the business model concerns its scalability, it is also a key related to its ability to adapt its resources and components by increasing or decreasing them in relation to the capabilities and needs of the market where the organization operates (Jabłoński 2015).

Source: A. Jabłoński, Usługi klimatyczne w kształtowaniu modeli biznesu przedsiębiorstw [Climate services in shaping the business models of companies], Przegląd Organizacji 4/2023, p. 145

dynamic constructs that change over time (Demil and Lecocq 2010, pp. 227–246). The business model differs from other units of analysis by its holistic perspective that spans organizational boundaries and different system levels (Zott et al. 2011, pp. 1019–1042). Business models are therefore difficult from an analytical, synthesizing, organizational and management perspective. Different definitions, approaches and interpretations related to the design and implementation of business model concepts occur. They arise from the individual context of each organization and the demands and expectations of the socio-economic ecosystem where they seek opportunities to capture economic value. The broad spectrum of definitions of business models results from their multidimensional application in strategic management processes and the ability to apply them effectively and efficiently. Such complexity causes a kind of chaos and uncertainty for managers in various fields of business in certain situations in the context of their appropriate selection in application terms. This situation also applies to the design and construction of business models that create positive climate change. In order to organize this area and develop the principles of functioning on the market, attention was paid to the types of business models that have a positive impact on climate change along with selected definitions. An appropriate dedicated summary is presented in the above Table 25.

The types of business models that positively influence climate change presented in the table above have different origins and different reasons for their design and operation, and are not only directly related to climate change. In many cases, they demonstrate a visible climatic impact, but this was not their direct goal when they were designed and created. Hence, there is undoubtedly a strong need to develop a targeted detailed definition of a climate-changing business model (CMB), dedicated to positive climate change what the Table 26 shows.

It should be noted that the author's proposed original definition of a climate-changing business model has a complex, precisely defined dimension. On the one hand, it refers to the creation of climate benefits and values for different groups of the

Table 26. A new type of Climate-Changing Business Model according to the author's original definition.

No.	A new type of business model with a positive impact on climate change	Author's original definition of a climate-changing business model
1.	Climate-Changing Business Model (CMB)	A climate-changing business model is a business model that aims to create climate benefits and values for different groups of the ecosystem of society, the economy, and business. The result of this model is usually climate-positive services generating a concurrent climate and economic impact. The platform for creating such business models is organizational culture shaped in organizations based on the assumptions of corporate climate responsibility.

Source: Jabłoński (2023, p. 146).

ecosystem of society, the economy and business. On the other hand, it is linked to the outcome of this model, which may be new, innovative climate services. The resulting character of this business model materializes in the form of a concurrent climate and economic impact. This has a proper market-based justification in today's reality; the platform for creating such business models is organizational culture shaped in organizations based on the assumptions of corporate climate responsibility. This approach enables group and individual behaviors embedded in the organizational culture to determine the sustainability of this business model in the short and long term. We can be tempted to say that in the long term, the climate-changing business model will permanently adapt to new economic and climate values, ensuring the organization's survival in even the most difficult market conditions. To link climate services to business models, it is important to use co-productions to build resilience to climate change. It is also important to track the progress of dynamic scientific research along with an analysis of the impact of these activities on all stages of the project cycle dedicated to given products or services. New business models should be dedicated to specifically implemented climate services and tracking and searching for specific schemes for financing climate action and developing climate-related investments in both the public and private sectors. In the context of verifying the value and benefits of such business models, the use of state-of-the-art seasonal and long-term climate forecasts becomes crucial. It is essential to have the ability to demonstrate the benefits and specific, quantifiable economic values of climate services to society and business. By delving into the issues of systemic construction of business models in terms of components along with the determination of optimal interfaces between individual components, their appropriate categorization becomes important. Analyzing the assumptions of the existence of business and the economy in relation to climate change, the aim was to determine the division and typology of individual groups of components which change the climate of business models. Four types were proposed in two groups, i.e., a general group and a specific group. The former included components of business models resulting from the general logic of doing business and components of business models resulting from the overall benefits offered and climate values. The latter included business model components resulting from specific conditions of a given sector of the economy neutralizing its negative impact on climate change as well as business model components resulting from specific conditions of a given sector of the economy positively affecting climate change. In many cases, they even fundamentally change the logic of its smart value proposal aimed at a positive impact on climate change, as shown in the table below.

For a complete and comprehensive picture of the position of a given component group in their application hierarchy, the pyramid of the component structure of climate-changing business models is presented in the Table 27. It is worth paying attention to the levels of location of individual component groups in relation to the general business logic and the specific groups of components dedicated to a given sector of the economy appropriately assigned and operated in the assumptions developed for business models. Pyramid of the component structure of climate-changing business models shows Figure 9.

Table 27. Typology of groups of components that change the climate of business models.

No.	Group of components changing the climate of business models	
1.	General Group	Business model components deriving from the general logic of doing business. This group of business model components includes classic business model components embedded in the business model canvas (Osterwalder and Pigneur 2010, p. 44).
2.		Business model components resulting from the overall climate benefits and values offered. This group of business model components can include components related to the overall climate benefits and values offered, which can include climate risk, green technologies, and carbon footprint in the value chain.
3.	Specific Group	Business model components resulting from the specific conditions of a given sector of the economy neutralizing its negative impact on climate change. This group of business model components may include components resulting from the specific conditions of a given sector of the economy neutralizing its negative impact on climate change. This concerns sector-specific measures to reduce emissions to air, emissions to water, emissions to soil, efficient waste management and the optimal use of resources.
4.		Business model components resulting from the specific conditions of a given sector of the economy, which positively influence climate change and in many cases even change the logic of the value proposition offered within aimed at a positive impact on climate change. This group of business model components can include components resulting from the specific conditions of a given sector of the economy positively affecting climate change and in many cases even changing the logic of the value proposition offered within aimed at a positive impact on climate change. In this case, the result of this group of components is to offer new climate services as a value proposition for customers and other stakeholders, expressed simultaneously in the climate and economic effect produced by the organization.

Source: Jabłoński (2023, p. 147)

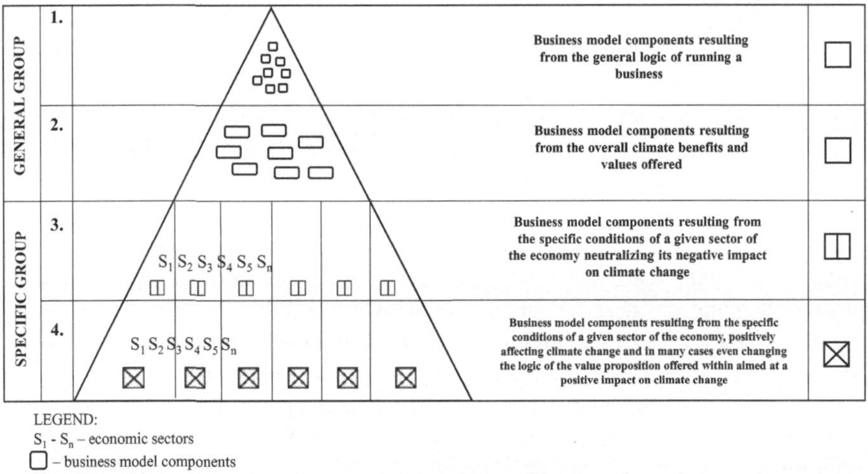

Figure 9. Pyramid of the component structure of climate-changing business models. Source: Jabłoński (2023, p. 147).

Based on the above scientific discussion, it is crucial to propose the author's original canvas of climate-changing business models along with its cause-and-effect dependencies. It includes previously discussed individual groups of business model components. The heart of this canvas includes the climate value proposition and the result of the implementation of these business models is the simultaneous achievement of the climate and economic impact by the organization. The proposed canvas of climate-changing business models is presented in the Figure 10.

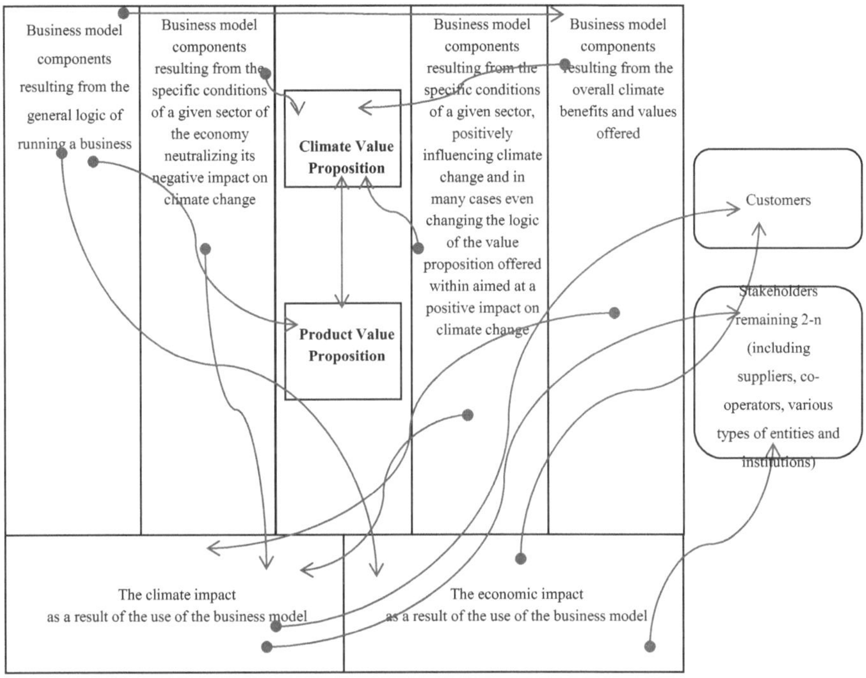

Figure 10. The proposed canvas of climate-changing business models. Source: Jabłoński (2023, p. 148).

The last step of the presented systemic methodology for the component-based construction of the optimal configuration of climate-changing business models is to link the type of climate service with a core business model component creating positive climate services for achieving the values and benefits generated by climate services for society and the economy. This results in the previously described consistency and compatibility of the offered climate services with the climate-changing business model implemented. This relationship is shown in the Table 28.

It seems that this proposed process of building the configuration of business models creating climate services in holistic terms should be looked at. This gives a greater guarantee of a broad understanding of climate conditions, their complex dimension and a strong impact on various areas of society, the economy and business. At the same time, it opens new cognitive possibilities and the creative creation of new business models and new climate services. Climate services used to shape the business models of companies are now becoming one of the most important factors in creating a climate-sensitive, responsible and modern economy. A new perspective

Table 28. Linking the type of climate service with a core component of business models that create positive climate services for achieving the values and benefits generated by climate services for society and the economy.

No.	Type of climate service offered	Group of climate-changing business model components	Values and benefits generated by climate services using climate-changing business models
1.	Information service on climate and climate change	Business model components for communication and positive climate relationships	Smart, climate-friendly information value
2.	A service that meets climate criteria and is rich in smart climate values	Business model components generating new services that turn negative climate impacts into positive ones	Smart, climate-friendly service value
3.	A product that meets the climate criteria and is rich in smart climate values	Business model components generating new products transforming negative climate effects into positive ones	Smart, climatic product value

Source: Own study.

on climate change and its strategic perspective as an opportunity rather than a problem determines new market opportunities while generating a positive impact on different areas of human life. The concept of business models is undoubtedly a tool that can contribute to strategic changes within climate threats in a multifaceted way. The transition from neutralizing the negative impact on the positive climate change created at the level of the global economy with sharp changes in sectoral conditions gives new perspectives for combating climate change. The specifically dedicated methodology and concept presented concern the component-based construction of the configurations of climate-changing business models with their interpretation in the specific context of the current ecosystem of society, the economy and business. Summarizing the in-depth scientific discussion in this area, the theoretical aspects of the author's original methodology and concept can include in particular:

- a multidimensional approach to climate services as a new way to increase the positive impact on climate change,
- linking climate services as a new climate value proposition with the concept of business models,
- developing a new typology of climate-changing business models,
- developing the logic and methodology for constructing climate-changing configurations of business models.

The utilitarian aspects of the developed methodology and concept include:

- introducing new organizational and technical solutions to various sectors of the economy through the climate services offered,
- constructing practical configurations that change the climate of business models,
- offering new climate services embedded in new business models of various types of organizations.

In view of the above inference, the significance of the problem being solved for economic theory and practice particularly concerns the development of theoretical assumptions for the creation of new business models and new, innovative climate services and their application through dedicated business models. Scientific research limitations related to the discussed issues are related to the new and innovative approach to the discussed areas of science and their wide cognitive complexity. This means that a number of topics related to climate change that definitely require further, in-depth scientific research and creative solutions to scientific problems in this important area of business, the economy and society emerge here.

8.4 Conclusions

The creation of new theoretical and practical space in the area of strategic management by means of business models can definitely change and reverse the negative polarization of the climate impact of companies. The economy undoubtedly has a huge impact on natural ecosystems. Companies should be fully aware of their environmental and climate impact and should be able to seek creative business ideas, including in current regulations and legislation. The creation of appropriate synergies in the area of climate adaptation can be a strong strategic component of the decision-making processes of climate-sensitive entrepreneurs and managers. This strategic horizon becomes a new and innovative space for the development and growth of climate-responsible companies seeking to build a new stream of activities of a long-term existence on the market.

8.5 Strategic recommendations in relation to the original methodology and concept of building climate-changing business models

Strategic recommendations for business and the economy resulting from the scientific discussion related to the development of the author's original methodology and concept of building climate-changing business models are as follows:

1. Adverse climate change can be stopped, neutralized or eliminated only through systemic organizational, management and technological solutions.
2. Decision-making processes related to climate change carried out by managers should be conducted by implementing business models that positively change the climate in many types of organizations.
3. Implementing the principles of climate responsibility can be an effective and efficient strategic instrument for creating new, innovative climate services.
4. Climate services should be provided through business models that result in a smart climate value proposal.
5. Climate-responsible organizations conduct their business processes with the awareness of simultaneously generating a consistent economic and climate impact.

References

Alexander, M. and Dessai, S. (2019). *What can climate services learn from the broader services literature?* Climatic Change 157: 133–149.

Asrar, G.R., Ryabinin, V. and Detemmerman, V. (2012). *Climate science and services: Providing climate information for adaptation, sustainable development and risk management*, Current Opinion in Environmental Sustainability 4(1): 88–100.

Baldissera Pacchetti, M. and St. Clair, A.L. (2023). *Framework to support the equitable standardisation of climate services*, D1.2 of the Climateurope project.

Bessembinder, J., Terrado, M., Hewitt, Ch., Garrett, N., Kotova, L., Buonocore, M. and Groenland, R. (2019). *Need for a common typology of climate services*, Climate Services, Vol. 16, art. 100135.

Bisaro, A., Hinkel, J., Le Cozannet, G., van der Pol, T. and Haas, A. (2021). *Global Climate Services: A Typology of Global Decisions Influenced by Climate Risk, Sec. Global Change and the Future Ocean*. Frontiers in Marine Science", Vol. 8.

Boon, E., Wright, S.J., Biesbroek, R., Goosen, H. and Ludwig, F. (2022). *Successful climate services for adaptation: What we know, don't know and need to know*, Climate Services, Vol. 27, art. 100314.

Brasseur, G.P. and Gallardo, L. (2016), *Climate services: lessons learned and future prospects*, Earth's Future 4(3): 79–89.

Brenner, T., Capasso, M., Duschl, M., Frenken, K. and Treibich, T. (2017). *Causal relations between knowledge-intensive business services and regional employment growth*, Regional Studies, 52(2): 172–183.

Climateurope. (2020), *Linking science and society, What do we mean by Climate Services*, https://www.climateurope.eu/definitions-climate-services/, accessed: 04.12.2023.

Demil, B. and Lecocq, X. (2010). *Business model evolution: in search of dynamic consistency. Long. Range Plan*. 43 (2e3): 227–246.

European Commission. (2015). *A European Research and Innovation Roadmap for Climate Services; Directorate-General for Research and Innovation (European Commission)*, Brussels, Belgium.

EU. (2020). *EU Strategy on Adaptation to Climate Change, European Parliament resolution of 17 December 2020 on the EU Strategy on Adaptation to Climate Change* (2020/2532(RSP)) (2021/C 445/19).

Grönroos, C. (1999). *Relationship marketing: challenges for the organisation*. Journal of Business Research 46(3): 327–335.

Hewitt, C., Mason, S. and Walland, D. (2012), *The global framework for climate services*. Nature Climate Change 2(12): 831–832.

Hewitt, Ch.D. and Stone, R. (2021). *Climate services for managing societal risks and opportunities*, Climate Services, Vol. 23, art. 100240.

Jabłoński, A. (2013a). *Modele zrównoważonego biznesu w budowie długoterminowej wartości przedsiębiorstw z uwzględnieniem ich społecznej odpowiedzialności* [Sustainable business models in building the long-term value of companies taking into account their social responsibility], Difin, Warszawa.

Jabłoński, A. (2013b). *Zielony wymiar budowy modeli biznesu i strategii przedsiębiorstw* [The green dimension of building business models and company strategies], [in:] Urszula Wąsikiewicz-Rusnak (Ed.) (collective work), *Współczesne problemy rozwoju przedsiębiorstwa w koncepcji rozwoju zrównoważonego* [Contemporary problems of company development in the concept of sustainable development*]*, WSB Dąbrowa Górnicza.

Jabłoński, A. (2015). *Design and Operationalization of Technological Business Models*, Acta Universitatis Agriculturae et Silviculturae Mendelianae Brunensis, Vol. 63, No. 3.

Jabłoński, A. (2023). Usługi klimatyczne w kształtowaniu modeli biznesu przedsiębiorstw [Climate services in shaping business models of companies], Przegląd Organizacji, 4/2023, p.144.

Jones, L., Carabine, E., Roux, J. and Tanner, T. (2015a). *Promoting the use of climate information to achieve long-term development objectives in sub-Saharan Africa: Results from the future climate for Africa scoping phase*. London, Climate and Development Knowledge Network.

Jones, L., Champalle, C., Chesterman, S., Cramer, L. and Crane, T.A. (2017). *Constraining and enabling factors to using long-term climate information in decision-making*. Climate Policy, 17(5): 551–572.

Jones, L., Dougill, A., Jones, R.G., Steynor, A., Watkiss, P., Kane, C., Koelle, B., Moufouma-Okia, W., Padgham, J., Ranger, N., Roux, J.P., Suarez, P., Tanner, T. and Vincent, K. (2015b). *Ensuring climate information guides long-term development.* Nature Climate Change 5: 812–814.

Larosa, F. and Perrels, A. (2017). *Assessment of the existing resourcing and quality assurance of current climate services, Grant agreement 730500, EU-MACS European Market for Climate Services*, Deliverable 1.2.

Larson, B.A. (2007). *Sustainable Development Research Advances*, New York, Nova Science Publishers.

Lipcaneanu, N. and Amanatidis, G. (2023). *Walka ze zmianą klimatu*, Parlament Europejski - Noty tematyczne o Unii Europejskiej, 09-2023, https://www.europarl.europa.eu/factsheets/pl/sheet/72/walka-ze-zmiana-klimatu, accessed: 04.12.2023.

Mentink, B. (2014). *Circular Business Model Innovation A process framework and a tool for business model innovation in a circular economy*. MSc thesis Industrial Ecology Master Theis. Delft University of Technology in Delft (Holland) and Leiden University in Leiden (Holland).

Ministry of Agriculture and Rural Development, Climate, Service of the Republic of Poland, https://www.gov.pl/web/agriculture/climate, accessed: 04.12.2023.

Osawa, Y. and Miyazaki, K. (2006), *An empirical analysis of the valley of death: large-scale R&D project performance in a Japanese diversified company.* Asian Journal of Technology Innovation 14(2): 93–116.

Osterwalder, A. and Pigneur, Y. (2010), *Business model generation: a handbook for visionaries, game changers, and challengers*, John Wiley & Sons, New York.

Rochlin, C. (2021), *Climate change and the progressive business model.* The Electricity Journal, Vol. 34, No. 4, art. 106927.

Rosenstock, T.S., Lubberink, R., Gondwe, S., Manyise, T. and Dentoni, D. (2020), *Inclusive and adaptive business models for climate-smart value creation.* Current Opinion in Environmental Sustainability 42: 76–81.

Tall, A. and Njinga, J. (2013), *Developing a methodology to evaluate climate services for farmers in Africa and South Asia workshop report*, CGIAR Program on Climate Change, Copenhagen, Denmark, Agriculture and Food Security, www.ccafs.cgiar.org.

Teece, D.J. (2010). *Business models, business strategy and innovation.* Long. Range Plan. 43: 172–194.

Tricco, A.C., Lillie, E., Zarin, W., O'Brien, K.K., Colquhoun, H., Levac, D., Moher, D., Peters, M.D.J., Horsley, T., Weeks, L., Hempel, S., Akl, E.A., Chang, Ch., McGowan, J., Stewart, L., Hartling, L., Aldcroft, A., Wilson, M.G., Garritty, Ch., Lewin, S., Godfrey, Ch.M., Macdonald, M.T., Langlois, E.V., Soares-Weiser, K., Moriarty, J., Clifford, T., Tunçalp, Ö. and Straus, S.E. (2018). *PRISMA extension for scoping reviews (PRISMA-ScR): Checklist and Explanation.* Annals of Internal Medicine 169(7): 467–473.

Vaughan, C., Dessai, S. and Hewitt, C. (2018). *Surveying climate services: what can we learn from a bird's eye view?* Weather, Climate, and Society 10(2): 373–395.

Visscher, K., Stegmaier, P., Damm, A., Hamaker-Taylor, R., Harjanne, A. and Giordano, R. (2020). *Matching supply and demand: A typology of climate services.* Climate Services, Vol. 17, art. 100136.

World Meteorological Organization. *Global Framework for Climate Services (GFCS)*, https://wmo.int/site/global-framework-climate-services-gfcs, dostęp: 04.12.2023.

Yunus, M., Moingeon, B. and Lehmann-Ortega, L. (2010), *Building social business models: lessons from the Grameen experience.* Long-range Planning, 43(2/3): 308–325.

Zott, C., Amit, R. and Massa, L. (2011). *The business model: recent developments and future research.* JOM 37: 1019–1042.

9

The Author's Original Methodology and the Concept of Shaping the Climate Maturity of the Business Models of Climate-Responsible Companies

9.1 Introduction

The purpose of this chapter is to present the author's original methodology and concept of shaping the climate maturity of the business models of climate-responsible companies. The scope of the chapter includes the research methodology adopted for shaping the climate maturity of the business models of climate-responsible companies. The subject of the chapter is to explain and create a proper understanding of the potential inherent in shaping the climate maturity of the business models of climate-responsible companies.

9.2 Methodological assumptions for the construction of the author's original methodology and concept of shaping the climate maturity of business models

The adopted mechanisms for shaping the climate responsibility of companies must and do result from a number of different factors and conditions, which can have an internal and external dimension. Regardless of their nature, the development of climate responsibility at the organizational level strongly depends on the awareness and climate sensitivity of companies embedded in its organizational genotype creating its organizational culture and, consequently, its climate culture. The result of the development of these attributes of the organization is wise management aimed at building platforms for generating decision-making processes that ensure the long-term existence of the organization. Such a platform can undoubtedly be the concept of shaping the climate maturity of the business models of companies. The concept of management maturity is one of the core components affecting the effectiveness and efficiency of organizational and management processes at all levels and functions

DOI: 10.1201/9781003474401-9

of management. The strategic management of the organization's maturity is vitally important here. Appropriate mechanisms for building the capacity to implement the organization's maturity model have been used for many years (Persse 2001). Hence, it is crucial to transfer the principles of building the maturity of the organization to the principles of building the strategic maturity of the business model and, in the context of climate responsibility, to the construction and logic and methodology of shaping the climate maturity of business models of companies. At this point, it is worth asking the key question: How can one build the concept of shaping the climate maturity of the business models of companies in an optimal way for the organization? The development of an original concept of shaping the climate maturity of the business models of companies is essential in the processes of a dynamic response to climate anomalies and dysfunctions, which, unfortunately, currently pose a huge threat both at the level of the organization itself in terms of management mechanisms and also in relation to the critical infrastructure of the organization in the context of ensuring its full and complete ability to carry out production or service processes. It should therefore be noted that organizations themselves operate at different levels of maturity. Measuring the maturity of an organization in the form of a maturity audit is one of the solutions that can be appropriately applied to achieve the appropriate and expected final results. This approach can also be used to shape the climate maturity of business models.

The original concept of shaping the climate maturity of the business models of companies presented herein is a critical and evaluating study based on current solutions related to the construction and development of business management systems from environmental management systems to climate management systems, taking into account the perspective of improving these systems, resulting from the business practices used so far, climate responsibility criteria at the level of companies and directions for their further development. The presented concept was implemented on the basis of a cognitive approach aimed at searching for the best possible economically optimal management mechanisms. Assumptions for the construction of business models based on the author's original concept of shaping the climate maturity of business models were implemented under the terms of the adopted methodology based on the construction, in the first step, of assumptions and, in the second step, of the construction of a reference model, understood as "a reference model for understanding significant relationships among the entities of some environment, and for the development of consistent standards or specifications supporting that environment" (OASIS SOA 2011). Reference models have emerged and are essential in Business Process Reengineering (BRP) (Champy 2009). The reference model is treated as a reference point that defines a coherent future solution in its entirety. It gives a sense of reduced risk and pursuit of an objectively optimal state. In creating the reference model, the best practice method was used, consisting of identifying solutions for the company's climate management system, as a set of complete guidelines and solutions facilitating the implication of guidelines resulting from development needs and principles for shaping the climate maturity of business models. The reference model used here is a hierarchical system, built according to the leading scheme of "strategy propagation top-down", where the symbols of responsibility, which can be reproduced in reality, in a specific organization are marked. In this approach, the reference model is treated as a mapping of the system configuration of climate maturity of the climate-responsible business model, which can serve to multiply real, concretely functioning organizations. This applies to both procedural and systemic aspects. When creating the reference model, vertical consistency was ensured, deriving from the decomposition of the core mechanisms of effectiveness and efficiency in the adopted climate criterion from the strategic level to the operational level. On this basis, a specific company will be able to adapt the proposed solutions and implement them in their own organizational structure.

The Figure 11 presents the concept of building a reference model and referring it to actual individual climate-responsible organizations.

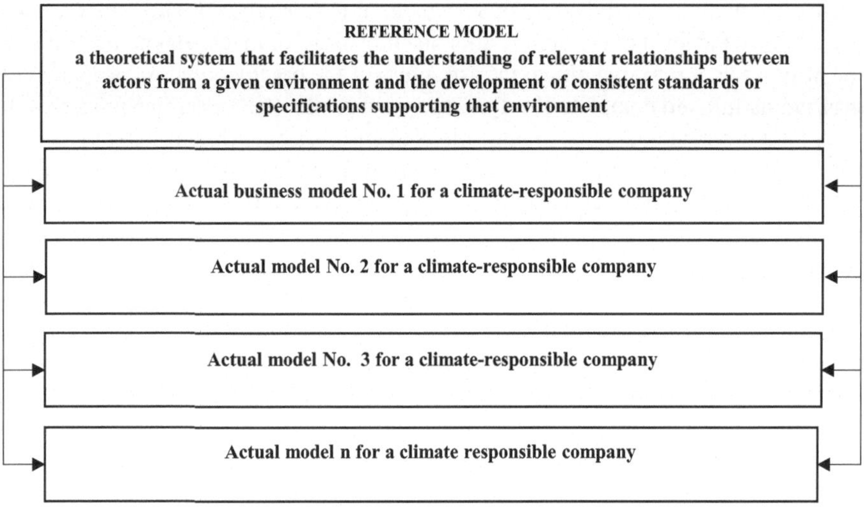

Figure 11. The concept of building a reference model for actual climate-responsible organizations. Source: Own study.

The proposed reference model aims to make a direct reference to the existing business models of companies in this case and their static configuration and their implementation in the dynamic dimension, assuming that a given organization has considered as part of its strategic behavior the fact that it wants to be and is a climate-responsible organization.

9.3 Criteria for building the author's original methodology and concept for shaping the climate maturity of business models (reference model)

The core assumption of the adopted methodology related to the development of the author's original concept of shaping the climate maturity of the business models of companies (a reference model) was the adoption of the following boundary conditions:

1. The adopted solution is a model solution, which includes a package of solutions directly serving a climate-responsible company.

2. Assuming that the proposed solution is of a model nature, it underlies the development of its own original structure for shaping the climate maturity of the business model of a specific climate-responsible company for a given organization along with appropriate emphasis on its own management mechanisms and its implementation in economic practice in a given market and an economic sector.
3. The users of this concept of shaping the climate maturity of the business models of companies should additionally draw inspiration for how to implement specific solutions in their company, including from other approaches, solutions and concepts in this area.

At this point, it should be noted that, as indicated by the practice of numerous companies around the world, current approaches regarding the construction of climate responsibility are, according to the author of the monograph, definitely insufficient. Therefore, the growing role and place of top management in shaping strategic and tactical–operational pro-climate behavior has been strongly emphasized. In order to provide a full package of tools to convince top management of the importance of the abovementioned issue of shaping the climate maturity of the business models of companies, the following boundary conditions, which were defined at the beginning of the development of this original concept and the idea of which guided the entire process of creating this monograph, are presented in the form of a diagnosis:

1. The concept adopted and proposed is the minimum requirement for a climate-responsible company and does not limit other climate activities of organizations using the principles of climate maturity of business models of companies.
2. The adopted concept contains a number of important solutions for direct application in this area.
3. In the course of developing the concept, the principles of a practical procedural and systemic approach were used.
4. The adopted original concept has been designed so that it can be implemented in every company around the world.
5. The assumptions for the development of the concept were based on the best solutions in the field of climate management based on the current state of scientific knowledge and management practice.
6. The proposed solutions are not contradictory and do not interfere with the solutions used so far in the field of building the climate maturity of business models of companies, but instead complement them and emphasize new interpretations that were not revealed by previous solutions.
7. Accepted business solutions implemented on the basis of their conceptualization should not be mistaken for the ways of operationalizing them in business practice. It is expected of users that, on the basis of the documentation provided, they will draw inspiration for solutions, adapting them to their business.
8. If this solution coincides with the solutions already used in the organization related to its climate organizational maturity, there is no need to start a new implementation procedure.

9. The analysis is voluntary and is not intended to be applied in full and uncritically without taking into account the own specificity of the functioning of the climate management system at the strategic and tactical and operational level in the organization, as well as the level of maturity of this system in the organization.
10. The intention of the original concept was not to demonstrate the redundancy and excessive complexity of the solution. It is purposeful to refer to different requirements and arrangements, even if the user covers these solutions with other solutions used in their company.
11. The proposed solutions indicate the directions and scope of further improvement of current solutions in the field of shaping the climate maturity of business models of climate-responsible companies.
12. A conscious user should be able to easily understand and implement the proposed solutions.
13. It should be assumed that the concept includes processes generating specific solutions using best practices available in the field of science and practice.

The logic and methodology related to the construction of the author's original concept of shaping the climate maturity of business models of companies (reference model) adopted and presented in this way has a strong theoretical basis, which can fully justify the understanding of the defined assumptions. It is worth mentioning that the adopted boundary conditions can be used to understand organizational and management requirements and form the basis for potential users to implement their own solutions in the area of corporate climate management.

9.4 Conceptualization and operationalization of the author's original methodology and concept of shaping the climate maturity of the business models of climate-responsible companies

The aim of the author's original concept of shaping the climate maturity of the business models of companies is to ensure that the company achieves its economic, long-term and short-term goals through its business models in a climate-responsible way. Such a strategic management system should be integrated with other management systems in order to improve the overall performance of the organization and reduce costs, while spreading efforts across all levels and functions of the organization. A climate-responsible company should have a developed and applied organizational structure, providing a division of tasks and responsibilities and a list and description of processes and projects, including tasks relevant to fulfilling climate criteria specific to the types of activities and services performed by the organization. In this sense, the proposed solution should include:

- the climatic context of the organization (and the appropriate management of this context),
- the climate leadership of the organization's top management (climate leadership and commitment, management of climate culture as a sub-part of the overall organizational culture of the company, climate management policy, the

development, supervision and management of programs for satisfying and enhancing climate criteria, the assessment of compliance with standards and legal requirements in the field of the climate and climate change, including the principles of climate adaptation of companies, functions, responsibilities and powers within the climate responsible organization, consultation with employees and other parties understanding the principles of climate responsibility),

- planning (climate risk planning and management, the identification and assessment of climate, process, functional and technical risks, climate risk management, strategic and tactical–operational climate goals),
- activity (the planning and supervision of operational activities related to climate management, the management of assets relevant to climate adaptation, contractors, partners and suppliers linked to our smart climate value chain, purchases meeting climate criteria, climate change management within our network of internal and external relations, crisis management related to emerging climate threats, and climate emergency management plans),
- support (resources focused on the organization's climate development, the organization's climate competency management system, training programs in the field of the climate and climate change, climate awareness and sensitivity, information and communication in the field of the climate and climate change, the preparation of a climate management system book, the management of human and organizational factors in the area of satisfying and improving climate criteria),
- an assessment of the company's climate performance (monitoring and measuring the climate and climate change, inspections to ensure compliance with climate criteria, an internal audit of compliance with climate criteria, a review of climate management and climate change at the level of the organization and its stakeholders and/or actors in the company's business network, the assessment of the impact of other management activities on compliance with the principles of climate responsibility), and
- improvement (drawing conclusions from climate events, continuous improvement in the area of shaping the organization's climate responsibility, corrective, preventive actions and supervision of non-compliances in the area of the climate and climate change).

The original concept of shaping the climate maturity of the business models of climate-responsible companies determines the resultant approach related to the concept of performance management. It is necessary to build a tight process, whereby all processes of the organization are measured by using dedicated indicators of KPI achievements in the area of the climate and climate change.

All processes, especially those related to satisfying climate criteria, are embedded in a dedicated map of processes aimed at shaping pro-climate behavior and fulfilling the principles of climate responsibility. A turtle diagram has been designed for them, where the course of the process is described, for which indicators of climate achievements have been defined and take into account the specificity of the functioning of a climate-responsible company. The model of the author's original concept of shaping the climate maturity of the business models of climate-responsible companies enables the constant monitoring of the climatic level of the organization through its systematic assessment. A turtle diagram is a qualitative tool used to help design or improve current organizational processes. The use of the turtle diagram helped obtain the best configuration of the process of shaping the climate maturity of the business models of climate-responsible companies. The use of the diagram for the design of core business model components, processes and projects has enabled the accurate and complete acquisition of the necessary data, information and knowledge and the creation of an economically optimal business model canvas built from the most tailored, adjusted business model components and the associated climate strategy and all necessary core processes and projects in line with the requirements of all stakeholders. Each element of the turtle diagram is important in terms of the principles of climate responsibility and has been described in detail and consciously according to the model below.

Input – detailed data entry for the process: documents, materials, plans, requirements, information, etc.

Process – the name of the process.

Output – information on what we obtain from the process: documents, materials, plans, requirements, information, etc.

What – what is necessary to complete the process: equipment, machinery, measuring equipment, computer system, software, etc.

Who – who is necessary for the smooth operation of the process: people and their competences, including the necessary training where competences are lacking.

How – anything that can help with "how" questions during the process: related processes, standards, instructions, procedures, visualizations, specific working methods, etc.

Criteria – process objectives and tasks, i.e., what we want to achieve and how we are able to measure it – KPI.

The chart of the climate management process aimed at shaping the principles of climate maturity, shown by means of the turtle diagram, is presented below in Figure 12.

In the main part of the process demonstrated by means of the turtle diagram, a model of the course of activities developing the principles of climate responsibility in the form of a flowchart is additionally developed. This is a tool used to present subsequent activities that fill in the adopted mechanisms of shaping the principles of climate maturity described by means of a turtle diagram using a block diagram. The aim of this procedure was to graphically present the process by means of geometric figures, which resulted in an easy-to-read and fully understood flow of information.

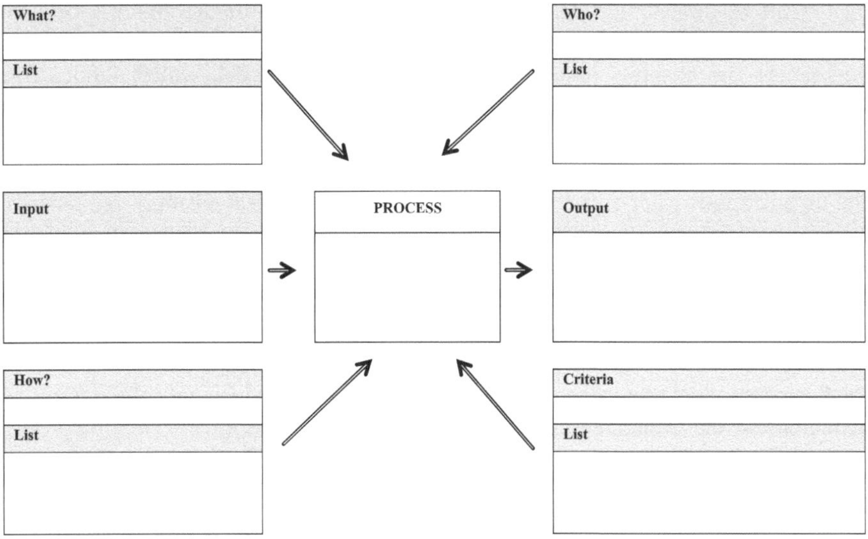

Figure 12. Chart of the climate management process aimed at shaping the principles of climate maturity, shown by means of a turtle diagram. Source: Own study.

9.5 The process of implementing the author's original methodology and concept of shaping the climate maturity of the business models of climate-responsible companies

The starting point for developing the author's original concept of shaping the climate maturity of the business models of climate-responsible companies, based on the principles of improvement of the organization in relation to its business model cooperating with the company's strategy and its processes and projects, should be reflected in directly and indirectly documented solutions used by the organization (e.g., the principles of a balanced scorecard focused on climate performance, a system of management by objectives focused on climate objectives, a climate risk management system, a personnel management system for the development of climate competence, a smart climate value management system or other management concepts, methods and techniques applicable to the adopted mechanisms for shaping the climate maturity of the company). Here, a business model management maturity model, which determines the degree of use of various types of solutions in relation to the current and shaped climate culture, is used in the context of offering a smart climate value proposal and income generation logic, where the genotype of building the principles of climate responsibility at the level of all components and functions of the business model of a climate-responsible company is embedded.

In the documentation describing the implementation of the author's original concept of shaping the climate maturity of the business models of climate-responsible companies, the business model management maturity model, which defines five levels of climate excellence of the company, was developed. This approach changes the point of view on the design and implementation of the company's business model, in the sense that currently a zero–one approach to fulfilling climate requirements by the organization is not sufficient. In particular, the organization must demonstrate an appropriate level of climate maturity in its business model. It is also crucial to increase the level of climatic maturity of the overall business management system. If the organization fulfills the core requirements specified in the business model management maturity model, it reaches the third level of climate system maturity. The fourth and fifth levels of maturity of the climate management system in the area of the business model require the implementation of new management solutions aimed at increasing the effectiveness of the climate management system. This approach completely changes the perspective of understanding the business model focused on climate criteria as well as preparations for periodic evaluations of this concept. This level of understanding and interpretation of scientific knowledge is absolutely crucial for fulfilling relevant and well-interpreted climate requirements, where the ability to develop an entire organization open to the mechanisms of climate change and adaptation in a strategic and tactical–operational approach and its improvement should be demonstrated.

The Figure 13 shows an example of a radar chart for visualizing multidimensional data in the climate maturity area, where each axis corresponds to one characteristic or indicator and the radius length corresponds to the value of that characteristic or indicator. This chart shows an example of the level of climate maturity of the business models of climate-responsible companies.

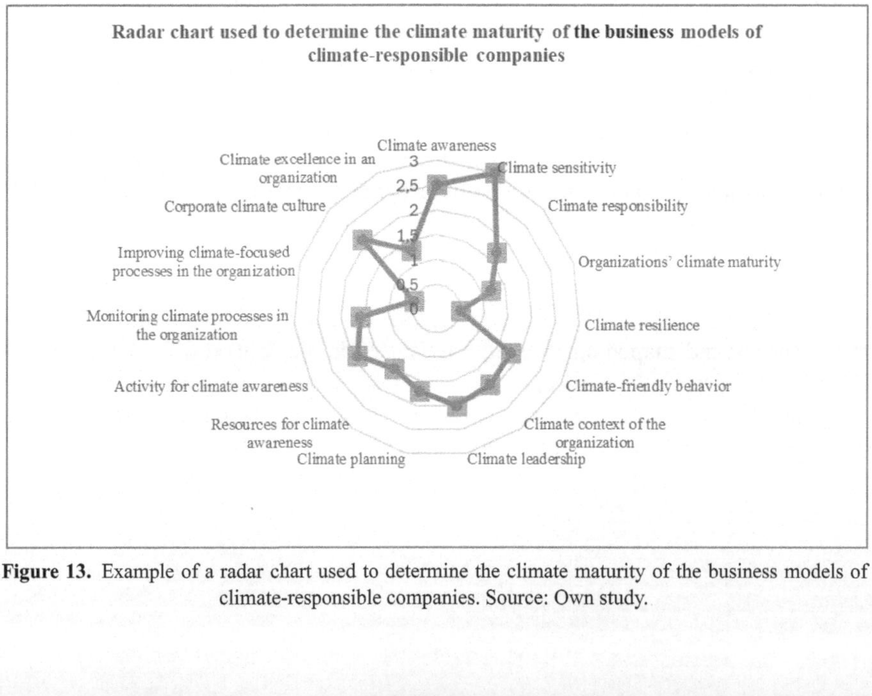

Figure 13. Example of a radar chart used to determine the climate maturity of the business models of climate-responsible companies. Source: Own study.

The presented approach and appropriate visualization of the resulting level of climate maturity of the business models of climate-responsible companies is fundamental to and critical for the overall development of the organization in a strategic and tactical–operational approach to the proper fulfillment of the principles of climate responsibility and the path to increasing the level of climate maturity of various types of organizations.

9.6 Strategic recommendations in relation to shaping the climate maturity of the business models of climate-responsible companies in the context of the development of digital technologies applicable to business models

Strategic recommendations for business and the economy resulting from the scientific discussion related to the development of the author's original methodology and concept of shaping the climate maturity of the business models of climate-responsible companies, in the context of the development of digital technologies applicable to the business models of companies, are as follows:

1. Determining the level of climate maturity of the business model and the measurement and monitoring thereof is a crucial element of the dynamic development of climate responsibility in various types of organizations.
2. Embedding the mechanisms of climate maturity in the organizational culture of companies opens a number of ways and opportunities to search for innovative solutions in the field of digitalization, technology and development of Artificial Intelligence at the level of climate-responsible companies.
3. The systemic and multi-criteria development of climate maturity, not only at the level of business models but also strategies, processes and projects, can generate the dynamic growth of business benefits while satisfying climate criteria and requirements.
4. The construction of a development platform of an organization based on the principles of climate maturity can be an excellent reason for the innovative shaping of the competitive advantage and competitive potential of companies open to positive climate change.

9.7 Conclusions

The application of the author's proposed original concept of shaping the climate maturity of the business models of climate-responsible companies may be significant for understanding the logic, methodology and conceptual framework related to finding and obtaining measurable economic, social and climate benefits in one properly interpreted coherence of various goals expected by the organization and stakeholders and/or actors of their business network gathered around them. This is followed by appropriate synergy, symmetry and symbiosis in the space of suitably applied decision-making systems using the design, implementation and operation of business models of climate-responsible companies.

References

Champy, J. (2009). *Reengineering Management*, HarperCollins e-books.
OASIS SOA Reference Model OASIS SOA Technical Committee. Retrieved 13/13/2011.
Persse, J.R. (2001). *Implementing the Capability Maturity Model*. New York: Wiley Computer Publishing.

10

The Author's Original Methodology and the Concept of Shaping the Climate Consistency of the Business Models of Climate-Responsible Companies

10.1 Introduction

The aim of this chapter is to present the author's original methodology and concept of shaping the climate cohesion of the business models of climate-responsible companies. The scope of the chapter includes the adopted research methodology for shaping the climate consistency of the business models of climate-responsible companies. The subject of the chapter is to clarify and create a proper understanding of the potential of shaping the climate consistency of the business models of climate-responsible companies.

10.2 Methodological assumptions for the development of the author's original methodology and the concept of shaping the climate consistency of the business models of climate-responsible companies

A company's business model is an appropriate, consistent and stable configuration of a group of core components aggregated in the business model, between which there are appropriate interfaces ensuring the maximization of the value proposition generated for customers as a result of the resulting value stream and the strong potential built into the business model. This business model is driven by the set of long-term goals of organizations embedded in the company's strategy itself. The consistency of the business model and strategy as well as the consistency between the business model and the strategy of the organization becomes crucial. In this specific case, the relationship between the business model and strategy is

DOI: 10.1201/9781003474401-10

dynamically mutual, i.e., the business model affects the company's strategy and vice versa, the strategy interacting within the framework of the feedback on the business model. In this space–time and from the broad perspective of scientific knowledge, this consistency becomes a guarantee of the strong use of the best possible solutions to achieve the appropriate effect in this area of climate management. At this point, therefore, it is also worth asking the question: How can one build the concept of the climate consistency of companies' business models and strategies in a way that is optimal for the organization? Below, a universal answer to the question is found.

The concept of consistency plays a very important role in the area of the development of management and quality science, especially in terms of strategic management mechanisms. This is crucial in the design, implementation and operation of companies' business models and strategies. Consistency binds elements together (OASIS SOA 2011). Consistency can be defined as a gradable feature of an organization consisting of such dependence and connection of all its components where a sufficiently strong change in any component may cause changes in the others. Therefore, the essence of consistency is the existence of relationships and connections between the components of the organization (Cyfert 2006, p. 15). Business model consistency is also understood as the ability of the business model to be fully configured to ensure its integrity, enabling the implementation of its functionality towards the achievement of high efficiency by the company (Jabłoński 2015, p. 75). Business model consistency is a whole, close and compatible connection of its components with each other and with the business environment, contributing to the creation of value for customers and capturing value for the company (Falencikowski 2013, p. 96). Business model consistency refers to the concept of business model matching, where this is to be used to achieve the objectives and results set for it by the model. Business model consistency is also related to the configuration of the business model, which determines the shape and image of the business model in not only static terms, but also dynamic terms. The configuration of the business model is a combination of properly selected components of the business model such that it ensures the achievement of the maximum possible (in a given situation, the internal and external environment of the company) increase in company value (Jabłoński 2013, p. 398). The configuration of the business model therefore constitutes its architecture in such a way that it can be used to capture the maximum value for the company. To describe the business model, the concept of total integrity is used, among others, which is understood as the absence of weakness in any component of the business model (Osterwalder and Pigneur 2010, p. 212). The concept of dynamic consistency also arises. Dynamic consistency is the ability to find a compromise between business model consistency (and thus efficiency) and a reality that, in most industries, will constantly change, constantly influencing and rejecting the dynamic movements inherent in the sequence, the trends of the business model and drifting with established actions aimed at consistently maintaining balance and direction

(Demil and Lecocq 2010, pp. 227–246). The consistency of business processes can be achieved through four types of operating models:

1. Diversification (poor standardization, poor integration).
2. Coordination (poor standardization, strong integration).
3. Replication (strong standardization, poor integration).
4. Unification (strong standardization, strong integration) (Ross et al. 2010, p. 53).

Business model consistency can be examined from the perspective of shaping climate responsibility. Then, an important attribute of shaping the principles of climate responsibility may be the development of the concept of shaping the climate consistency of the business models of climate-responsible companies.

10.3 Conceptualization and operationalization of the author's original methodology and the concept of shaping the climate consistency of the business models of climate-responsible companies

In order to be able to achieve an appropriate climate impact at the level of the company and its relationships with its market environment, the mechanisms of conceptualization and operationalization of the author's original methodology and concept of shaping the climate consistency of the business models of climate-responsible companies can be implemented through the four levels of climate consistency proposed below.

Level I - Climatic Market Consistency

Level II - Internal Component Consistency of the Business Model in Its Canvas

Level III - Strategic Climate Consistency

Level IV - Tactical and Operational Climate Consistency

The implementation of the principles of proposed consistency should be a continuous process aimed at cascading these levels from level I to level IV. The Table 29 provides a list of the levels of climate consistency and their descriptions.

The proposed logic of cascading levels of climate consistency within an organization is very important for its specific effect and result, which should be generated by maximizing the internal and external climate value. Then, a wide stream of positive impacts on the climate and its adaptation is created, measured by the combined economic, climatic, social and ecological value achieved for the benefit of organizations and society.

Table 29. List and description of levels of climate consistency.

No.	Level of climate consistency	Type of climate consistency	Description of attributes of climate consistency
1.	Level I Climatic Market Consistency	Company – Customers	The purpose of the business model is to capture value from the market by offering a value proposition to customers for which they are willing to pay and indeed do so. Thus, the result of generating this value proposition is the cash received from these customers. In the relationship between consistency and the market, the attribute of this consistency is the price for the products and/or services offered, which customers accept in this case. Consistency then occurs between the value proposition offered to customers and the price that customers pay for that proposition. The price is composed of the cost of producing that value proposition and the organization's margin on that value proposition. Market climate consistency will then occur between the company and customers if this value proposition contains climate criteria which are not only accepted by customers but, above all, the climate criteria embedded in the value proposition are the sources of the choice of the value proposition. Therefore, if smart climate value is included in this value proposition for climate-sensitive customers and provides a platform for their purchasing decision-making processes, we can talk about climate consistency in the organization–customer relationship.
2.	Level II Internal Component Consistency of the Business Model in Its Canvas	Business model components – Business model components	In order to be able to create optimal solutions related to the conceptualization and operationalization of the author's original methodology and the concept of shaping the climate consistency of the business models of climate-responsible companies, it is necessary to refer to the component structure of business models. A business model in this sense is considered an aggregated set of components in the designated function of its purpose. With regard to the component structure of the company's business model, climate consistency will occur when business model components that positively affect the climate and its changes are included in the canvas of this business model, and the resultant canvas will be a climate-smart value proposition for customers who will pay for it and wish to pay for it on a continuous and systematic basis because they have sufficient awareness and climate sensitivity.
3.	Level III Strategic Climate Consistency	Business model – Strategy	In order to be able to effectively and efficiently implement and operate the business model of a company, it is crucial to maintain strategic consistency between the business model and strategy. If the business model in its internal component consistency is a climate-changing business model, it will be put into motion through a climate strategy that is an aggregated set of long-term strategic climate goals.
4.	Level IV Climate Tactical and Operational Consistency	Processes – Projects	The processes and projects of the company are at the service of the business model and strategy of the organization. If the business model and strategy are focused on offering the climate value proposition and implementing the climate strategy, these goals are implemented through climate-compliant processes and projects at tactical and operational levels. The solutions used therein generate assumptions of climate responsibility.

Source: Own study.

10.4 Strategic recommendations with regard to shaping the climate consistency of the business models of climate-responsible companies in the context of the development of digital technologies applicable to business models

Strategic recommendations for business and the economy resulting from the undertaken scientific discussion related to the development of an original methodology and the concept of shaping the climate consistency of the business models of climate-responsible companies in the context of the development of digital technologies applicable to business models are as follows:

1. The mechanisms adopted for cascading the levels of climate consistency should be relevant to the state of development of the digital economy and society.
2. The climate consistency solutions applied at the level of the company–customer market relationship should include digital marketing mechanisms in addition to analog solutions.
3. The solutions of climate consistency applied at the level of the component structure of the business model should take into account the principles of cybernetics as a science of control applicable to adjusting the internal business model consistency.
4. The climate consistency solutions applied at the level of the business model–strategy relationship should take into account digital communication and the development of this relationship in accordance with the adopted principles of digital and climate transformation in the long term.
5. The climate consistency solutions applied at the level of relationships between the project processes should, inter alia, apply the principles of the circular, low- and zero-emission economy and eliminate the carbon footprint and digital carbon footprint.

10.5 Conclusions

The adopted solutions regarding the conceptualization and operationalization of shaping the climate consistency of the business models of climate-responsible companies are vitally important in the processes of cost optimization while maximizing the climate impact. Through the maintenance of consistency criteria, actions are structured and based on the principles of dynamic equilibrium, ensuring that climate impact is achieved without unnecessary and uncontrolled costs and unnecessary, unjustified and chaotic actions. This approach is aimed at a clear decision-making process based on the principles of usability and business rationality.

References

Cyfert, S. (2006). *Strategiczne doskonalenie architektury procesów w zarządzaniu przedsiębiorstwem* [Strategic improvement of the process architecture in business management]. Wydawnictwo Akademii Ekonomicznej w Poznaniu. Poznań, p. 15.

Falencikowski, T. (2013). *Spójność modeli biznesu, koncepcja i pomiar* [Consistency of business models, concept and measurement]. CeDeWu, Warszawa, p. 96.

Osterwalder, A. and Pigneur, Y. (2010). *Business Model Generation*. John Wiley & Sons, Inc., New Jersey, USA, p. 212.

Ross, J.W., Weil, P. and Robertson, D.C. (2010). *Architektura korporacyjna jako strategia* [Enterprise architecture as strategy], Studio Emka, Warszawa, p. 53.

11

The Author's Original Methodology and the Concept of Shaping Climate Responsibility in the Context of Designing Business Models from the Perspective of Management, Digital Transformation and the Technological Development of Companies

11.1 Introduction

The aim of this chapter is to present the author's original methodology and the concept of shaping climate responsibility in the context of designing business models from the perspective of management, digital transformation and the technological development of companies. The scope of the chapter includes the adopted research methodology for shaping climate responsibility in the context of designing business models from the perspective of management, digital transformation and the technological development of companies. The subject of the chapter is to explain and create a proper understanding of the potential of shaping climate responsibility in the context of designing business models from the perspective of management, digital transformation and the technological development of companies.

DOI: 10.1201/9781003474401-11

11.2 Methodological assumptions for building the author's original methodology and the concept of shaping climate responsibility in the context of designing business models from the perspective of management, digital transformation and the technological development of companies

The development of the author's original methodology and the concept of shaping climate responsibility in the context of designing business models from the perspective of management, digital transformation and the technological development of companies is a very important element of shaping the stream of positive changes in the area of climate change at the level of companies, the economy and society. This methodology should give rise to even the most difficult and perhaps unpopular decision-making processes at different levels of management and in different bodies of climate decision-makers in both national and international dimensions. All this requires an appropriate emotional and substantive commitment to achieving difficult goals and the related results. These results should be visible to all stakeholders and/or actors of the various international and local entities and communities involved in the qualitative development of climate change.

11.3 Conceptualization and operationalization of the author's original methodology and the concept of shaping climate responsibility in the context of designing business models from the perspective of management, digital transformation and the technological development of companies

The conceptualization and operationalization of the author's original methodology and the concept of shaping climate responsibility in the context of designing business models from the perspective of management, digital transformation and the technological development of companies was focused on embedding core climate attributes in the most important ontological entities of the organization, which include processes, projects, business models and strategies. In this view, the following attributes of climate responsibility were created:

1. The attribute of climate process responsibility
2. The attribute of climate project responsibility
3. The attribute of climate responsibility embedded in the logical structure of the business model
4. The attribute of climate responsibility embedded in the stream of long-term climate objectives

The core attributes listed in the Table 30 describe the principles of climate responsibility in order to adopt an appropriate approach to understanding, applying

Table 30. Core attributes describing the principles of climate responsibility in order to adopt an appropriate approach to understanding, applying and creating the principles of climate responsibility in the context of the development of digital technologies applicable to the key ontological entities of companies.

No.	Ontological Entity	Attribute of Climate Responsibility	Description of the Attribute of Climate Responsibility
1.	Processes	The attribute of climate process responsibility in a climate value chain focused on circular economy principles and the elimination of the carbon footprint and digital carbon footprint	The use of the attribute of climate process responsibility is of organizational–management, technical–operational and economic–financial importance because building climate responsibility in processes gives a multidimensional climate impact. It covers the entire, complete value chain open to building smart climate value in a way that reduces negative production and service results. All possible solutions that determine the achievement of the assumed positive climate impact with respect to the principles of generating a reasonable profit and maintaining the complete financial liquidity of the organization are used. Optimization mechanisms are built into all processes of the organization.
2.	Projects	The attribute of climate project responsibility in the individual, unique and innovative implementation of a stream of well-thought-out project activities composed of activities meeting the climate criteria	The implementation of the attribute of climate project responsibility in the company is crucial when we understand a company as the sum of ongoing, implemented projects in the appointed function of time. Thus understood, a design company open to generating positive climate value builds appropriate climatic criteria for a given project.
3.	Business models	The attribute of climate responsibility embedded in the logical structure of the business model	A climate-responsible business model is a business model that is a unique, individualized configuration of the components contained therein, where they themselves and the interfaces between these components create a proposal for smart climate value for sensitive and climate-conscious customers. The revenue-generating logic embedded in this business model includes fulfilling climate criteria throughout its entire structure.
4.	Strategies	The attribute of climate responsibility embedded in the stream of long-term climate objectives	The climate strategy fulfills its role by pursuing long-term climate goals that ensure the organization's ability to survive even in the toughest market conditions by seeking value and meeting them for all climate-sensitive stakeholders of the organization and/or the actors of its business network.

Source: Own study.

and creating the principles of climate responsibility in the context of the development of digital technologies applicable to the key ontological entities of companies.

The attributes presented and their descriptions are fundamental in a broad and complex approach to shaping the principles of climate responsibility with reference to companies for which requirements and climate criteria are important. It should also be noted that the comprehensive coverage of the range of decision-making processes in the organization offers an amazing space for the development of the climate competences of the organization's staff and the shaping of climate sensitivity and awareness among them.

11.4 Strategic recommendations with regard to shaping the climate responsibility of the business models of climate-responsible companies in the context of the development of digital technologies applicable to the business models of companies

Strategic recommendations for business and the economy resulting from the undertaken scientific discussion related to the development of the author's original methodology and the concept of shaping the climate responsibility of the business models of climate-responsible companies in the context of the development of digital technologies applicable to the business models of companies are as follows:

1. The climate responsibility of processes should be focused on the search for all possible solutions in order to eliminate adverse climate impact by opening up to positive climate adaptation, especially digital and technological solutions.
2. Climate responsibility at the project level should be embedded in standardization mechanisms and dedicated project management methodologies.
3. Climate responsibility at the level of business models should be open to changing the polarization of creating the value proposition to offer smart climate value and thus dynamically changing the behavior of strategic customers and, above all, to seeing climate components in products and services and determining innovative and creative climate products and services.
4. The climate responsibility of the strategy should pave the way for new market opportunities in the potential for adaptation to climate change. At the same time, it should be aimed at a new type of customer, the so-called climate customers, i.e., those customers who are guided by climate criteria and requirements in their purchasing processes.

11.5 Conclusions

The adopted mechanisms and principles included in the conceptualization and operationalization of the author's original methodology and the concept of shaping climate responsibility in the context of business model design should provide a platform for long-term and conscious management primarily aimed at the future of the organization. Short-term, temporary and short-sighted thinking when making decisions that do not take into account climate criteria is not a guarantee of survival and, as a consequence, can offer quick short-term profits and other benefits, but from the point of view of the logic and methodology of a broad and spatial understanding of economic, social, ecological and climate needs, is not, in fact, justified.

12

Development of a Set of Organizational–Management, Technical–Operational and Economic–Financial Factors with the Greatest Impact on Climate Responsibility in the Context of Designing Business Models from the Perspective of Management, Digital Transformation and the Technological Development of Companies

12.1 Introduction

The aim of the chapter is to present a set of organizational–management, technical–operational and economic–financial factors with the greatest impact on climate responsibility in the context of designing business models from the perspective of management, digital transformation and the technological development of companies.

The scope of the chapter covers the organizational, technical and financial components in relation to the assumptions of climate responsibility.

The subject of the chapter concerns the design of business models shaping the mechanisms of climate responsibility.

DOI: 10.1201/9781003474401-12

12.2 Core factors influencing climate responsibility in the context of business model design

In order to be able to properly and logically understand, interpret and apply the principles of climate responsibility in a comprehensive way, it is important to develop a set of organizational, management, technical, operational, economic and financial factors which have the greatest impact on climate responsibility in the context of designing business models from the perspective of management, digital transformation and the technological development of companies, as presented below.

The set of organizational and management factors with the greatest impact on shaping climate responsibility includes:

1. Building the right mechanisms for building business models that positively change the climate.
2. Building the right mechanisms for building adaptation strategies that positively change the climate.
3. Building a consistent strategic hybrid that positively changes the climate, composed of interrelated processes, projects, business models and strategies.
4. Defining the core attributes of shaping climate responsibility at the strategic, tactical and operational levels of the organization.
5. Building optimal mechanisms for shaping the organizational culture and the organizational climate for the benefit of the climate by developing the climate culture in organizations.
6. Adapting the organizational structure and core functions of the organization to the principles of developing and promoting climate responsibility in the organization.
7. Implementing the principles of the systemic and network approach in building a positive relationship with stakeholders and/or network actors for the implementation of the principles of climate responsibility.
8. Building the core attributes of building an organization's smart climate value.
9. Building the core attributes of shaping the climate maturity of business models of climate-responsible companies.
10. Building the core attributes for shaping the climate consistency of the business models of climate-responsible companies.
11. Building the core attributes of building climate competence among the organization's staff and key stakeholders and/or actors of the organization's business network.

A set of technical and operational factors with the greatest impact on the formation of climate responsibility includes:

1. Transition from linear processes to circular approaches.
2. Transition to the low-carbon and carbon-free economy while implementing co-production principles.
3. Transition from analog models to digital climate processes and projects.

4. Developing Artificial Intelligence solutions in all possible applications in the organization.
5. Developing solutions for automation, robotics, technical diagnostics, mechatronics and technical and social cybernetics at the level of the entire organization.

A set of economic and financial factors with the greatest impact on shaping climate responsibility includes:

1. Transition from classical cost accounting to marginal cost accounting applicable in digital transformation.
2. Application of principles optimizing the organization's operating costs through the implementation of modern logistics systems along with the elimination of the carbon footprint and digital carbon footprint.
3. Implementation of modern methods of monetization of digital business models together with the principles of micropayments.
4. Development of the principles of financial climate risk in the organization in economic and financial terms.
5. Focusing on climate insurance processes to protect against negative climate events.

12.3 Conclusions

The presented list of organizational–management, technical–operational and economic–financial factors which have the greatest impact on climate responsibility in the context of designing business models from the perspective of management, digital transformation and the technological development of companies is above all very important in the application processes related to the implementation of a comprehensive system of shaping climate responsibility in various types of organizations. It should be remembered that almost every organization which is conscious of wise management always tries to search for systemic solutions that give the possibility of the permanent application of practical solutions, with the expected business, social and, in this case, primarily climate impact. This approach provides a number of market opportunities in relation to building platforms for the effectiveness and efficiency of the organization in the short and long term.

13

Discussion of Research Results

Conclusions for Theory, Practice, Future Research Directions and the Limitations of the Research Process, Theoretical Implications, Practical Implications, and Research Limitations

13.1 Introduction

The aim of the chapter is to present future research directions and limitations of the research process and theoretical implications, practical implications and potentially interesting directions for future research, taking into account the limitations of scientific research. The scope of the chapter covers the core elements of the scientific discourse on the development of corporate climate responsibility. The subject of the chapter concerns important elements of shaping climate responsibility in terms of future scientific research.

13.2 Scientific discourse on future research related to corporate climate responsibility

Theory constructively compared with practice is a panacea for numerous problems that require comprehensive solutions, including in the area of climate management. The combined implementation and cooperation of science with business practitioners can provide the world with appropriate tools, techniques and concepts in the area of business models and strategies for adaptation to climate change. Such a discourse requires positive, reciprocal intentions and a strong imperative to generate common added value for the climate and the resulting streams of strategic goals for the entire world.

DOI: 10.1201/9781003474401-13

The following are the assumptions for a multi-criteria and multi-faceted discussion on the results of scientific research related to the need to develop climate responsibility.

Conclusions for theory

1. Changing current management paradigms in the area of digital and climate transformation towards the development of new concepts of climate management.
2. Dynamic development of Artificial Intelligence aspects in theoretical terms in the context of digital and climate transformation.
3. Developing theoretical aspects of Industry 4.0 and Industry 5.0 in relation to climate management mechanisms.
4. Expanding the theoretical potential of the issue of climate responsibility in the context of the maturity and climate consistency of the business models of climate-responsible companies.
5. Comprehensive theoretical analysis of the place and role of digital climate technologies in shaping climate responsibility.
6. Comprehensive theoretical analysis of the construction and implementation of business models which positively change the climate through their appropriate attributes.
7. Theoretical link between the principles of sustainable development and business and climate development.
8. Developing mechanisms for shaping climate competences in organizations and their climate capital.
9. Building strong theoretical bridges between the scientific discipline of management and quality science and climate science.

Conclusions for practice

1. Implementing the principles of climate and digital transformation in all parts of the organization.
2. Dynamic development of regions, cities and companies through the implementation of the principles of climate responsibility.
3. Developing the principles of the scientific understanding of climate across the economy, business and society.
4. Developing new ways and solutions for climate forecasting taking into account Artificial Intelligence.
5. Focusing the principles of technological management on creating a positive climate impact.
6. Focusing on logistics management with the elimination of the carbon footprint and digital carbon footprint.
7. Developing production and service processes based on a low- and zero-carbon economy.
8. Developing the principles of co-production in organizational processes.

Future research directions and limitations of the research process in the area of climate management

1. Developing new climate services and products that change consumer perceptions and climate behavior.
2. Developing and exploring climate marketing concepts, including digital climate marketing.
3. Conducting experimental research in the area of new business models which positively affect the climate.
4. Developing modern principles of social communication for eliminating infodemics, disinformation and destabilization in the area of climate change.
5. Implementing comprehensive research, including comparative studies on the mechanisms of strategic change of sectors of the economy from those with a negative impact to those which positively affect the climate.
6. Developing research aimed at combining application solutions for the triad of Digital transformation – Climate transformation – Energy transformation.
7. Seeking optimal solutions for climate change in relation to the war in Ukraine and other armed conflicts from a military perspective.

13.3 Conclusions

The presented conclusions for theory, practice, future research directions and limitations of the research process, theoretical implications, practical implications, potentially interesting directions for future research, and the limitations of research are the essence of scientific discussion. Such a broad approach to climate issues provides a strong arsenal to be used by organizations to search for new space for scientific exploration aimed at the development of a practical process of managing organizations that are open to creating positive climate change.

14

Recommendations and Strategic Axioms in the Field of Climate Responsibility in the Context of Designing Business Models from the Perspective of Management, Digital Transformation and the Technological Development of Companies

14.1 Introduction

The aim of the chapter is to present strategic recommendations and axioms in the field of climate responsibility in the context of designing business models from the perspective of management, digital transformation and the technological development of companies. The scope of the chapter includes a set of strategic recommendations and axioms in the field of climate responsibility related to the design of business models. The subject of the chapter concerns core strategic recommendations and axioms in the field of climate responsibility related to the design of business models.

14.2 A set of core recommendations and strategic axioms in the field of climate responsibility related to the design of business models

As part of the research and scientific analyses, an analytical and evaluating study was developed, the conclusions of which allowed for the creation of a set of axioms proving

DOI: 10.1201/9781003474401-14

the intensity of theoretical and practical knowledge in the field of a solved scientific problem related to the issue of climate responsibility in the context of designing business models from the perspective of management, digital transformation and the technological development of companies. Axioms are statements isolated from all the theorems of a given theory, selected so that all the other theorems of that theory can be derived therefrom. This arrangement of axioms is called axiomatics. The set of axioms and their consequences is an axiomatic system.

There are different sets of recommendations for action in the area of climate change. Core strategic recommendations resulting from the adopted thought process contributing to the creation of positive climate values through digital transformation solutions include the following:

1. Climate change should be seen as an opportunity to create new business models as well as new climate products and services.
2. Entrepreneurs should treat climate change as a platform for new adaptation in an economically and socially difficult reality.
3. Climate change should make entrepreneurs aware that new products and services should be in line with society's expectations aimed at improving the quality of life in a climate crisis.
4. Artificial intelligence and automation and robotics solutions supported by the digitalization of the economy should be a powerful tool with which to generate positive climate change.
5. The climate transformation should be consistent and scalable with the digital transformation.
6. The combined achievement of climate and digital objectives should be a source of climate change strategies at the level of specific companies.
7. The implementation of business processes by companies should not focus on reducing the negative impact on the environment and climate, but instead become a platform for creating new climate-rich products and services.
8. Climate services should not only focus on information about the state of the environment and climate, but should be extended to products and services that enhance the attributes of the new climate-changing economy.
9. Corporate social responsibility should include a new area in the activity of companies focused on corporate climate responsibility.
10. Business monetization supported by micropayments using digital platforms should be strongly linked to a positive climate impact.
11. Digital products should responsibly displace material products that directly interfere with the natural environment and consequently have an adverse impact on the climate.
12. The resilience of business models should be linked to the achievement of climate resilience of individual regions of the world.
13. The stream of long-term strategic goals of the organization should be embedded in the realities of the climate crisis.

14. The creation of climate innovation should be a core activity of companies familiar with the climate realities.
15. A climate-friendly business model with a positive impact on the natural ecosystem should create a climate value proposition for a wide range of stakeholders and should be based on an income generation logic based on marginal costs.
16. Products and services that have a negative impact on the environment and the climate should be gradually converted into products that have a positive impact on the climate, while correcting the social behavior of a wide range of stakeholders.
17. Climate risk mitigation should be embedded in business risk management mechanisms of modern-minded companies.
18. Managers' decision-making processes should include micro-decisions, micro-targets and micro-corrections in the area of positive climate impact.
19. Automation and robotics supported by artificial intelligence solutions should be aimed at creating a circular model of production management, especially among large industrial companies.
20. The categorization of companies should include the concept of a climate company, i.e., such a company whose operating logic built into the business model, strategy, processes and projects should be focused on having a positive impact on the climate in all fields of the functioning of this organization. The resulting climate strategic hybrid built from the business model, strategy, processes and projects should be a source of new dynamics of management of companies generating climate value streams.
21. Company space management and its dynamics, creating a new dimension of companies in motion, should be aimed at shaping climate value in all possible dimensions of business and social existence.

The adaptation of companies to the climate crisis should be a key priority for modern strategically thinking companies (Orlove 2022). Climate priorities that generate a positive impact on the climate should be embedded in business models that adhere to the climate values of companies. The navigation and control of the company should include the creation of new products and services implemented in line with the assumptions of the circular economy supported by digital transformation solutions reducing the costs of the organization's operation. Building a climate company as a new ontological entity embedded in the theory and practice of today's business should open a new space for creating a consistent platform for achieving digital, climate, ecological, economic and social goals. The logic of the climate company embedded in its business model and strategy should shape economic and social changes at the level of various actors of the business ecosystem with a strong impact on creating space for generating positive climate change. The path of creating space for climate change should be supported by the legal dimension of shaping social and economic reality at both local and global levels.

14.3 Conclusions

The described recommendations and strategic axioms in the field of climate responsibility in the context of designing business models from the perspective of management, digital transformation and the technological development of companies offer the opportunity to apply them primarily to the development of organizations open to new areas of shaping competitive advantages by generating smart climate values embedded in the processes of shaping climate responsibility. The path of conscious management requires being open to new knowledge, different interpretations and new innovative applications thereof at the company level.

References

Orlove, B.S. (2022). The Concept of Adaptation. Annual Review of Environment and Resources. doi: 10.1146/annurev-environ-112320-095719.

Conclusion

The presented scientific monograph, in the opinion of its author, is significant in terms of the processes of the proper and multidimensional understanding of the principles of climate responsibility in business models, in relation to management, digital and technological perspectives. It should be noted that this issue currently has a particularly strong dimension of creative and positive strategic, complex and interdisciplinary thinking about the climate, as a number of core dimensions and areas of the discussion overlap with the solution to numerous current climate problems at the level of business, the economy and society, sometimes even understood and interpreted differently by various stakeholder groups and bodies at the political, social and economic levels. At the same time, this creates a great number of wrong and false interpretations, including those related to misinformation about real climate change, its current anomalies and dysfunctions. Therefore, the professional and systemic arrangement of scientific knowledge in relation to the development of the principles of climate responsibility at the strategic and tactical–operational levels of many types of organizations seems completely expedient and justified. The aim of the monograph is to provide, primarily in a scientific and theoretical manner, but also strongly focused on its constructive comparison with the practice of specialized management, a reliable and logically presented theoretical and practical trajectory for the search for the positive impact of various types of organizations on climate change and the creation of their adaptive potential. It should also be noted that the monograph attempts to show the positive aspect of climate change management primarily in the context of numerous approaches and dimensions and thus reverse the negative polarization of the understanding of climate change among companies from negative perception and orientation to positive perception and orientation. Companies, as the subjects of creative business thinking, should clearly see the emerging array of market opportunities in even the most difficult economic and social problems. Positive climate visionaries at the level of various types of companies should create new, innovative climate services and products by applying climate-changing business models and climate strategies, among other things. The negative impact on the climate should be firmly and decisively halted all over the world. In addition, the mere neutralization of this negative impact on the climate is also no longer sufficient. It is therefore necessary to seek positive and creative climate solutions in the short and long term without justifying these actions with inaction and powerlessness. After all, creative solutions should be expected and demanded from managers and entrepreneurs open to multidimensional and multicriteria creation and shaping their market success through the dynamic creation of space for the development of climate responsibility at all levels and management functions of various types of organizations.

DOI: 10.1201/9781003474401-15

Index